Lecture Notes in Mathematics

Edited by A. Dold and B. Eckmann

912

Numerical Analysis

Proceedings of the 9th Biennial Conference
Held at Dundee, Scotland, June 23–26, 1981

Edited by G. A. Watson

Springer-Verlag
Berlin Heidelberg New York 1982

Editor

G. Alistair Watson
Department of Mathematical Sciences, University of Dundee
Dundee DD1 4HN, Scotland

AMS Subject Classifications (1980): 65–06, 65D10, 65F99, 65K05, 65M25, 65N30, 65R20.

ISBN 3-540-11199-9 Springer-Verlag Berlin Heidelberg New York
ISBN 0-387-11199-9 Springer-Verlag New York Heidelberg Berlin

Printing and binding: Beltz Offsetdruck, Hemsbach/Bergstr.
2141/3140-543210

Dedicated to A. R. Mitchell

on the occasion of his 60th birthday

Preface

For the four days June 23 - 26, 1981, around 180 people from 22 countries gathered in Dundee, Scotland for the 9th Biennial Conference on Numerical Analysis held at the University of Dundee. Talks at the meeting were given by 15 invited speakers, and their papers appear in these notes. In addition to the invited papers, 63 shorter contributed talks were presented: a complete list of these, together with authors' addresses, is also given here.

This preface is the main opportunity which I have to thank all the speakers, including the after-dinner speaker at the conference dinner, Principal Adam Neville of the University of Dundee, all chairmen and participants for their contributions. I would also like to thank the many people in the Department of Mathematical Sciences of this University who assisted in various ways with the preparation for, and running of, this conference. In particular, the secretaries in the Department deserve special praise for the way in which they coped with the large volume of typing necessary in the few weeks leading up to the conference.

Financial support for this meeting was obtained from the European Research Office of the United States Army. In times of escalating costs, it would not be possible to offer a reasonably low-cost conference without such financial help, and this support is gratefully acknowledged. The conference is also indebted to the host University for the provision of a sherry reception for all participants, and also for making available various University facilities throughout the week.

There can be very few numerical analysts who do not associate with the University of Dundee the name A. R. Mitchell. He has had, and continues to have, a profound influence on numerical analysis, both in this country and beyond, and if the Dundee numerical analysis conferences have any claim to have become important events in the numerical analysis calendar, this can largely be attributed to his guidance, his enthusiasm and his personal magnetism. On the 22 June 1981, the day before the start of the conference, Ron Mitchell celebrated his 60th birthday; as a small token of the esteem in which he is held by the numerical analysis community, it was unanimously agreed by the conference that these proceedings should be dedicated to him on this occasion.

Dundee, November 1981 G. A. Watson

INVITED SPEAKERS

A R Conn:	Department of Computer Science, University of Waterloo, Waterloo, Ontario, Canada N2L 3G1.
G Dahlquist:	Department of Numerical Analysis and Computing Science, Royal Institute of Technology, S-100 44 Stockholm 70, Sweden.
L M Delves:	Department of Computational and Statistical Science, University of Liverpool, P O Box 147, Liverpool L69 3BX.
J Douglas, Jr:	Department of Mathematics, University of Chicago, 5734 University Avenue, Chicago, Illinois 60637, USA.
I S Duff:	Computer Science and Systems Division, AERE Harwell, Oxfordshire OX11 0RA, England.
R Fletcher:	Department of Mathematical Sciences, University of Dundee, Dundee DD1 4HN, Scotland.
C W Gear:	Department of Computer Science, University of Illinois at Urbana-Champaign, Urbana, Illinois 61801, USA.
G H Golub:	Department of Computer Science, Stanford University, Stanford, California 94305, USA.
J G Hayes:	Division of Numerical Analysis and Computer Science, National Physical Laboratory, Teddington, Middlesex TW11 0LW, England.
P J van der Houwen:	Mathematische Centrum, Kruislaan 413, 1098 SJ Amsterdam, The Netherlands.
T E Hull:	Department of Computer Science, University of Toronto, Toronto, Canada M5S 1A7.
P Lancaster:	Department of Mathematics and Statistics, University of Calgary, 2920 24 Ave. N.W., Calgary, Canada T2N 1N4.
R J Y McLeod:	Department of Mathematical Sciences, New Mexico State University, Las Cruces, New Mexico 88003, USA.
M J D Powell:	Department of Applied Mathematics and Theoretical Physics, University of Cambridge, Silver Street, Cambridge CB3 9EW, England.
L B Wahlbin:	Department of Mathematics, Cornell University, Ithaca, New York 14853, USA.

SUBMITTED PAPERS

M J Baines, Department of Mathematics, University of Reading, Reading, England.
An increment type monotonicity - preserving algorithm for conservation laws in several dimensions.

C T H Baker, Department of Mathematics, University of Manchester, Manchester M13 9PL, England.
Initial-value problems for integro-differential equations.

A Bellen, Institute of Mathematics, University of Trieste, Trieste, Italy.
Monotone methods for periodic solutions of second order differential equations.

M Berzins and P M Dew, Department of Computer Studies, University of Leeds, Leeds LS2 9JT, England.
A generalised Chebyshev method for non-linear parabolic equations in one space variable.

N R C Birkett and N K Nichols, Department of Mathematics, University of Reading, Reading, England.
Optimal control for oscillating systems with applications to energy extraction problems.

Erich Bohl, Faculty of Mathematics, University of Konstanz, Postfach 5560, 7750 Konstanz, W Germany.
On the numerical treatment of ordinary reaction-convection-diffusion models.

Klaus Böhmer, Department of Mathematics, University of Marburg, 3550 Marburg, W Germany.
A mesh independence principle for discretizations of operator equations.

N G Brown* and R Wait+, *Department of Building and Civil Engineering, Liverpool Polytechnic, Liverpool, England, +Department of Computational and Statistical Science, University of Liverpool, Liverpool, England.
A branching envelope reducing algorithm for finite element meshes.

James R Bunch, Department of Mathematics, University of California, San Diego, La Jolla, California 92093, USA.
Stable decomposition of skew-symmetric matrices.

I T Cameron, Department of Chemical Engineering and Chemical Technology, Imperial College, London SW7 2BY, England.
Numerical solution of differential-algebraic systems in chemical process dynamics.

Ole Caprani and Kaj Madsen, Institute of Datalogy, University of Copenhagen, DK-2200 Copenhagen, Denmark.
Error bounds for the solution of integral equations.

I D Coope and S P J Matthews, Department of Mathematics, University of Canterbury, Christchurch, New Zealand.
A convergent algorithm for linear discrete L_p approximation.

F Crowet and C Dierieck, MBLE, Philips Research Laboratory, 2 - Box 8, B 1170, Brussels, Belgium.
Streamfunction representation of an incompressible plane flow.

G Dahlquist, O Nevanlinna and W Liniger*, *IBM T J Watson Research Center, P O Box 218, Yorktown Heights, NY 10598, USA.
Unconditionally stable one-leg methods with variable steps.

A Davey, Department of Mathematics, University of Newcastle upon Tyne, NE1 7RU, England.
On the numerical solution of 'stiff' boundary value problems.

Nira Dyn and David Levin, School of Mathematical Science, Tel-Aviv University, Tel-Aviv, Israel.
A procedure for generating diagonal dominance in ill-conditioned systems originating from integral equations and surface interpolation.

G H Elliott, Department of Mathematics and Statistics, Portsmouth Polytechnic, Portsmouth PO1 2EG, England.
Polynomial approximation in the complex plane using generalised Humbert polynomials.

G Fairweather and R L Johnston, Department of Computer Science, University of Toronto, Toronto, Canada M5S 1A7.
Boundary methods for the numerical solution of problems in potential theory.

Hartmut Foerster, G.M.D., Postfach 1240, D-5205 St Augustin 1, W Germany.
On multigrid software for elliptic problems.

David M Gay, Center for Computational Research in Economics and Management Science, Massachusetts Institute of Technology, Cambridge, MA 02139, USA.
Solving interval linear equations.

L Grandinetti, Dipartimento di Sistemi, University of Calabria, 87036 Arcavacata (Cosenza), Italy.
Computational analysis of a new algorithm for nonlinear optimization which uses a conic model of the objective function.

J A Grant, Department of Mathematics, Bradford University, Bradford BD7 1DP, England.
On finding the zeros of a linear combination of Chebyshev polynomials.

J de G Gribble, Department of Mathematics, Dalhousie University, N.S., Canada.
Inner product quadrature formulas.

M K Horn, DFVLR-Oberpfaffenhofen, 8031 Wessling/Obb., W Germany.
High-order, scaled Runge-Kutta algorithms.

Ulrich Hornung, Institute for Numerical Mathematics, University of Münster, D-4400 Münster, W Germany.
Convergence of the transversal line method for a parabolic-elliptic equation.

A Iserles, Department of Applied Mathematics and Theoretical Physics, University of Cambridge, Silver Street, Cambridge CB3 9EW, England.
Numerical solution of linear ODE's with variable coefficients.

W Jureidini, Department of Mathematics, American University of Beirut, Beirut, Lebanon.
A numerical treatment of the stationary Navier-Stokes equations using nonconforming finite elements.

Pat Keast and Graeme Fairweather, Computer Science Department, University of Toronto, Toronto, Canada M5S 1A7.
On the H^{-1}-Galerkin method for second-order linear two-point boundary value problems.

Per Erik Koch, Institute of Information, Blindern, Oslo 3, Norway.
A trigonometric collocation method for two-point boundary value problems.

Fred T Krogh and Kris Stewart, Jet Propulsion Laboratory, California Institute of Technology, 4800 Oak Grove Drive, Pasadena, CA 91109, USA.
Implementation of variable step BDF methods for stiff ODE's.

Kuo Pen-Yu, Department of Mathematics, Shanghai University of Science and Technology, Shanghai, China.
On stability of discretization.

Sylvie Lescure, Direction des Etudes et Recherches, Electricite de France, 1 Avenue du Général de Gaulle, 92141 Clamart, France.
An energy criterion for crack propagation.

Matti Mäkelä, Department of Computer Science, University of Helsinki, SF-00250, Helsinki 25, Finland.
On the possibility of avoiding implicit methods and Jacobians when solving stiff equations.

J C Mason, Mathematics Branch, Royal Military College of Science, Shrivenham, Swindon, England.
Complex interpolation and approximation on an annulus, and applications in fracture mechanics.

S McKee, Computing Laboratory, University of Oxford, 19 Parks Road, Oxford, England.
The University Consortium for Industrial Numerical Analysis (UCINA).

G Meinardus, Department of Mathematics, University of Mannheim, 6800 Mannheim 1, W Germany.
Asymptotic behaviour of iteration sequences.

J P Milaszewicz, Department of Mathematics, Ciudad Universitaria, 1428 Buenos Aires, Argentina.
On modified Gauss-Seidel iterations.

N K Mooljee, Edinburgh Regional Computing Centre, Edinburgh EH9 3JZ, Scotland.
Some aspects of curve-fitting in a university environment.

A Murli and M A Pirozzi, Department of Mathematics, University of Naples, Naples, Italy.
Use of fast direct methods for mildly nonlinear elliptic difference equations.

Igor Najfeld, Institute for Computer Applications in Science and Engineering, NASA Langley Research Center, Hampton, VA 23665, USA.
Analytical-numerical solution of the matrix ODE $Y^{(k)} = AY + F(t)$.

S Nakazawa, Department of Chemical Engineering, University College of Swansea, Swansea SA2 8PP, Wales.
Some remarks on the 'upwind' finite elements.

P Onumanyi and E L Ortiz, Imperial College, University of London, London, England.
Correction, singular nonlinear boundary value problems and the recursive formulation of the Tau method.

Gerhard Opfer and Bodo Werner, Department of Mathematics, University of Hamburg, D-2000 Hamburg 13, W Germany.
Nonconforming complex planar splines.

Michael L Overton, Courant Institute, 251 Mercer Street, New York, NY 10012, USA.
A quadratically convergent method for minimizing a sum of Euclidean norms.

M H C Paardekooper, Department of Econometrics, Tilburg University, 225 Tilburg, The Netherlands.
Upper and lower bounds in aposteriori error analysis by Newton-Kantorovich techniques.

Herman J J te Riele, Mathematical Centre, Amsterdam, The Netherlands.
Collocation methods for weakly singular second kind Volterra integral equations.

L Rolfes and J A Snyman*, *University of Pretoria, Pretoria, South Africa.
A global method for solving stiff differential equations.

J M Sanz-Serna* and I Christie, *Department of Mathematics, University of Valladolid, Valladolid, Spain.
Product approximation in nonlinear eigenvalue problems.

R W H Sargent, Department of Chemical Engineering and Chemical Technology, Imperial College, London SW7 2BY, England.
Nonlinear programming algorithms and global, superlinear convergence.

D S Scott and R C Ward, Computer Sciences Division, Union Carbide Corporation - Nuclear Division, Oak Ridge, Tennessee 37830, USA.
Algorithms for sparse symmetric-definite quadratic λ-matrix eigenproblems.

L F Shampine, Applied Mathematics Research Department, Sandia National Laboratories, Albuquerque, New Mexico 87185, USA.
Recent progress in the automatic recognition of stiffness.

S Sigurdsson, Faculty of Engineering and Science, University of Iceland, Rejkjavik, Iceland.
A Galerkin procedure for estimating normal gradients in two-dimensional boundary value problems.

R D Skeel, Department of Mathematics, University of Manchester, Manchester M13 9PL, England.
Odd/even reduction for the adaptive solution of one dimensional parabolic PDEs.

D Sloan, Department of Mathematics, University of Strathclyde, Glasgow, Scotland.
Stability and accuracy of a class of numerical boundary conditions for the advection equation.

J A Snyman, University of Pretoria, Pretoria, South Africa.
A new and dynamic method for unconstrained minimization.

Per Grove Thomsen and Niels Houbak, Institute for Numerical Analysis, Technical University of Denmark, DK 2800, Lyngby, Denmark.
Iterative refinement techniques in the solution of large stiff systems of ODE's.

M van Veldhuizen, Department of Mathematics, Vrije University, The Netherlands.
Collocation on Gaussian abscissae for a singularly perturbed model problem.

Jan Verwer, Mathematical Centre, Amsterdam, The Netherlands.
Boundedness properties of Runge-Kutta-Rosenbrock methods.

H A Watts, Applied Mathematics Division 2646, Sandia National Laboratories, Albuquerque, New Mexico 87185, USA.
Computing eigenvalues of boundary value problems.

F C P Whitworth, Mathematics Department, Brighton Polytechnic, Brighton, England.
Global error estimates for general linear methods for ordinary differential equations.

P H M Wolkenfelt, Mathematical Centre, Amsterdam, The Netherlands.
Linear multilag methods for Volterra integral equations.

H Wolkowicz, Department of Mathematics, University of Alberta, Edmonton, Alberta, Canada T6G 2G1.
Solving unstable convex programs.

T J Ypma, Department of Applied Mathematics, University of the Witwatersrand, Johannesburg, 2001, South Africa.
How to find a multiple zero.

CONTENTS

A SECOND-ORDER METHOD FOR SOLVING THE CONTINUOUS MULTIFACILITY LOCATION PROBLEM†

P.H. CALAMAI and A.R. CONN

ABSTRACT

A unified and numerically stable second-order approach to the continuous multifacility location problem is presented. Although details are initially given for only the unconstrained Euclidean norm problem, we show how the framework can be readily extended to l_p norm and mixed norm problems as well as to constrained problems.

Since the objective function being considered is not everywhere differentiable the straightforward application of classical solution procedures is infeasible. The method presented is an extension of an earlier first-order technique of the authors and is based on certain non-orthogonal projections. For efficiency the linear substructures that are inherent in the problem are exploited in the implementation of the basic algorithm and in the manner of handling degeneracies and near-degeneracies. The line search we describe also makes use of the structure and properties of the problem. Moreover, the advantages that we derive from the linear substructures are equally applicable to small-scale and large-scale problems.

Some preliminary numerical results and comparisons are included.

1. INTRODUCTION

Since the 17th century, when Fermat first posed a single facility location problem involving Euclidean distances, the issue of locating an object according to some set of rules and criteria has recieved a great deal of attention in the literature. The bibliographies of Lea [11] and Francis and Goldstein [7] together represent well over a thousand references to these problems.

In general, location problems ask where some object or objects should be placed to improve a measure of the performance of the system in which they interact. Here, we consider a prototype location problem: the static and deterministic formulation of the minisum multifacility location problem involving l_p distances. The objective involves locating a set of new facilities (objects) in a system of existing facilities to minimize the sum of weighted l_p distances between the new and existing facilities and among the new facilities.

One difficulty with the l_p distance problem is that the objective function is not everywhere differentiable. In fact, nondifferentiability occurs whenever any two facilities coincide. The straightforward use of gradient reducing procedures to solve this problem is therefore inapplicable. However, various methods that circumvent this nondifferentiability have been used. For example, linear programming methods (see [15] and [21]) and gradient reduction methods on approximating functions (see [20]) have been used to solve the rectilinear distance (l_1) location problems. For Euclidean distance (l_2) problems, modified gradient reducing methods have also been used (see [6]

---

† This work was supported in part by Natural Science and Engineering Research Council of Canada Grant No. A8639.

and [12]), as have subgradient methods, pseudo-gradient methods and heuristic methods (see [3], [1] and [19]).

In this paper we present a projected Newton method for solving the l_p distance location problem and we describe an implementation of this method that takes full advantage of the structure of the problem and its graphic interpretation. This second-order technique is a natural extension of the first-order projected steepest descent algorithm reported in [2]. A similar extension, developed independently by Overton, is presented in [17]. In his paper, the quadratic convergence of a projected Newton method is proved and a special line search is described.

2. PROBLEM STATEMENT

The multifacility minisum problem involving costs associated with Euclidean distances between facilities in $\mathbf{R}^q$ can be stated as: Find the point $x^{*T} = \{x^{*T}_1, \ldots, x^{*T}_n\}$ in $\mathbf{R}^{qn}$ to minimize

$$f(x) = \sum_{1 \leqslant j < k \leqslant n} v_{jk} \| x_j - x_k \| + \sum_{j=1}^{n} \sum_{i=1}^{m} w_{ji} \| x_j - p_i \| , \tag{2.1}$$

where

$n \triangleq$ number of new facilities (NF's) to be located.

$m \triangleq$ number of existing facilities (EF's).

$x_j^T = (x_{j1} \cdots x_{jq}) \triangleq$ vector location of NF_j in $\mathbf{R}^q$, $j = 1, \ldots, n$.

$p_i^T = (p_{i1} \cdots p_{iq}) \triangleq$ vector location of EF_i in $\mathbf{R}^q$, $i = 1, \ldots, m$.

$v_{jk} \triangleq$ nonnegative constant of proportionality relating the l_2 distance between NF_j and NF_k to the cost incurred, $1 \leqslant j < k \leqslant n$.

$w_{ji} \triangleq$ nonnegative constant of proportionality relating the l_2 distance between NF_j and EF_i to the cost incurred, $1 \leqslant j \leqslant n$, $1 \leqslant i \leqslant m$.

$\| x_j - x_k \| = (\sum_{c=1}^{q} | x_{jc} - x_{kc} |^2)^{1/2} \triangleq l_2$ distance between NF_j and NF_k, $1 \leqslant j < k \leqslant n$.

$\| x_j - p_i \| = (\sum_{c=1}^{q} | x_{jc} - p_{ic} |^2)^{1/2} \triangleq l_2$ distance between NF_j and EF_i, $1 \leqslant j \leqslant n$, $1 \leqslant i \leqslant m$.

If we randomly collect all the nonzero v_{jk} constants into an ordered set where the first member is called α_1 and the last member is called α_η and then randomly collect all the nonzero w_{ji} constants into an ordered set where the first member is called $\alpha_{\eta+1}$ and the last member is called α_τ, then, if we define the index set $M = \{1, \ldots, \tau\}$, problem (2.1) can be restated more conveniently as

$$\text{minimize} \qquad f(x) = \sum_{i \in M} \| A_i^T x - b_i \| \tag{2.2}$$

where the $q \times qn$ matrix A_i^T and the $q \times 1$ vector b_i are defined by

$$A_i^T = [0_1 \quad \alpha_i I \quad 0_2 \quad \overline{\alpha}_i I \quad 0_3] \qquad i = 1, \ldots, \tau$$

$$b_i = (\alpha_i + \overline{\alpha}_i) p_{i^*} \qquad i = 1, \ldots, \tau$$

and

$$\overline{\alpha}_i = \begin{cases} -\alpha_i & i = 1, \ldots, \eta \\ 0 & i = \eta+1, \ldots, \tau \end{cases}$$

I is a $q \times q$ identity matrix

0_1 is a zero matrix of dimension $q \times q\,(j^* - 1)$

0_2 is a zero matrix of dimension $q \times q\,(k^* - j^* - 1)$

0_3 is a zero matrix of dimension $q \times q\,(n - k^*)$

$j^* = j$ where $\alpha_i \equiv v_{jk}$ or $\alpha_i \equiv w_{ji}$

$$k^* = \begin{cases} k & \text{where } \alpha_i \equiv v_{jk} \\ j+1 & \text{where } \alpha_i \equiv w_{ji} \end{cases}$$

$$i^* = \begin{cases} \text{an arbitrary constant} & \text{where } \alpha_i \equiv v_{jk} \\ i & \text{where } \alpha_i \equiv w_{ji} \end{cases}$$

3. ANALYSIS

3.1 Introduction

One difficulty in solving problem (2.2) arises because the convex objective function $f(x)$ is not everywhere differentiable. If we let

$$r_i(x) = A_i^T x - b_i \qquad \forall i \in M$$

then the objective function can be written as

$$f(x) = \sum_{i \in M \setminus I_\epsilon(x)} \| r_i(x) \| + \sum_{i \in I_\epsilon(x)} \| r_i(x) \| \tag{3.1.1}$$

$$= \tilde{f}(x) + \sum_{i \in I_\epsilon(x)} \| r_i(x) \| .$$

With a proper choice of the index set $I_\epsilon(x)$ we can guarantee that the function $\tilde{f}(x)$ is clearly differentiable in the neighbourhood of x and that the remaining expression contains all the nondifferentiable (and near-nondifferentiable) terms.

3.2 A First-Order Method

Suppose we wish to minimize the first-order change in the objective function $f(x)$ by moving in some direction h. This can often be accomplished by minimizing the unit first-order change in the function $\tilde{f}(x)$ subject to the condition that the first-order change in the remaining terms $\| r_i(x) \|$, $i \in I_\epsilon(x)$ remain zero. That is, we solve

$$\begin{array}{lll} \underset{h}{\text{minimize}} & h^T \nabla \tilde{f}(x) & \\ \text{subject to} & A_i^T h = 0 & i \in I_\epsilon(x) \end{array} \tag{3.2.1}$$

(Notice that we are able to take advantage of the linear substructure of this problem in the constraint terms.)

The solution to this problem yields the direction

$$h = -\gamma P \nabla \tilde{f}(x) \tag{3.2.2}$$

where P is the orthogonal projector onto $S^\perp$, S is the space spanned by the columns of A_i $i \in I_\epsilon(x)$ and $\gamma > 0$ is chosen to satisfy the bound on the norm of h. Thus, for $h = -\gamma P \nabla \tilde{f}(x) \neq 0$, $g = \nabla f$ and λ sufficiently small, we have

$$\begin{aligned} f(x+\lambda h) - f(x) &= \lambda g^T h + 0(\lambda^2) \\ &= -\lambda\gamma \| P \nabla \tilde{f}(x) \| + 0(\lambda^2) \\ &< 0 \; . \end{aligned} \tag{3.2.3}$$

If $P \nabla \tilde{f}(x) = 0$ then $\nabla \tilde{f}(x)$ must lie entirely in S (in this case we call the point x a dead point; see § 3.4). Letting $\mathbf{A} = [\overline{A}_{i_1} \cdots \overline{A}_{i_t}]$ where $I_\epsilon(x) = \{i_1 \cdots i_t\}$ and $\alpha_i \overline{A}_i = A_i$ $i \in I_\epsilon(x)$ then, assuming $\mathbf{A}$ is full rank, $\nabla \tilde{f}(x)$ can be uniquely expressed as

$$\begin{aligned} \nabla \tilde{f}(x) &= \mathbf{A} u \qquad u^T = [u_{i_1}{}^T \cdots u_{i_t}{}^T] \\ &= \sum_{i \in I_\epsilon(x)} \overline{A}_i u_i . \end{aligned} \tag{3.2.4}$$

(The vector u is called the Lagrange or dual vector; see § 3.4.) Then, for any choice of h, we have

$$\begin{aligned} f(x+\lambda h) - f(x) &= \lambda \left[h^T \nabla \tilde{f}(x) + \sum_{i \in I_0} \| A_i^T h \| \right. \\ &\quad \left. + \sum_{i \in I_1} h^T \nabla (\| r_i(x) \|) \right] + 0(\lambda^2) \\ &= \lambda \left[\sum_{i \in I_0} \left(u_i^T \overline{A}_i^T h + \alpha_i \| \overline{A}_i^T h \| \right) \right. \\ &\quad \left. + \sum_{i \in I_1} \left(u_i^T \overline{A}_i^T h + \alpha_i \frac{r_i(x)^T \overline{A}_i^T h}{\| r_i(x) \|} \right) \right] + 0(\lambda^2) \end{aligned} \tag{3.2.5}$$

where $I_0 = \{ i \in I_\epsilon(x) \mid \| r_i(x) \| = 0 \}$ and $I_1 = I_\epsilon(x) \backslash I_0$. If, under these circumstances, there exists an index $l \in I_\epsilon(x)$ such that $\| u_l \| > \alpha_l$ we take as our descent direction

$$\tilde{h}_l = -\gamma P_l \overline{A}_l u_l \tag{3.2.6}$$

where P_l is the orthogonal projector onto $S_l^\perp$, S_l is the space spanned by the columns of $\mathbf{A}$ with columns $\overline{A}_l$ deleted and $\overline{A}_l^T \tilde{h}_l = -\rho u_l$ where $\rho > 0$ (see [2]). For this choice of direction (i.e. $h = \tilde{h}_l$) and for sufficiently small $\lambda > 0$ we will have

$$f(x+\lambda h)-f(x) = \lambda g^T h + 0(\lambda^2)$$

$$= \begin{cases} -\lambda\rho\left[\|u_l\|^2 - \alpha_l\|u_l\|\right] + 0(\lambda^2) & l \in I_0 \\ -\lambda\rho\left[\|u_l\|^2 + \alpha_l\dfrac{r_l(x)^T u_l}{\|r_l(x)\|}\right] + 0(\lambda^2) & l \in I_1 \end{cases} \tag{3.2.7}$$

$$< 0$$

where

$$g = \begin{cases} \nabla\tilde{f}(x) - \dfrac{A_l u_l}{\|u_l\|} & l \in I_0 \\ \nabla\tilde{f}(x) + \dfrac{A_l r_l(x)}{\|r_l(x)\|} & l \in I_1 \end{cases} .$$

For detail of this first-order method see [2].

3.3 A Second-Order Method

We wish now to find a direction h which minimizes the second-order change in the objective function. This can be accomplished by minimizing the change in the function $f(x)$ up to second-order terms subject to the condition that the change in the remaining expressions remain zero up to second-order terms. Thus, we solve the following problem

$$\underset{h}{\text{minimize}} \; h^T\nabla\tilde{f}(x) + \tfrac{1}{2}h^T\nabla^2\tilde{f}h \tag{3.3.1}$$

$$\text{subject to} \quad A_i^T h = 0 \quad i \in I_\epsilon(x).$$

(The quadratic programming problem with quadratic constraints that ensues in [4] when both function and constraint curvature are included has been simplified because of the linear substructure of this problem's constraints.)

Now, define a $qn \times q(n-t)$ matrix Z satisfying

$$A_i^T Z = 0 \qquad i \in I_\epsilon(x) \tag{3.3.2}$$

$$Z^T Z = I_{q(n-t)} \tag{3.3.3}$$

and the transformation $h = Zw$ so that problem (3.3.1) becomes

$$\min_{w} w^T Z^T \nabla\tilde{f} + 1/2\, w^T Z^T \nabla^2\tilde{f} Z w \tag{3.3.4}$$

The solution to this problem can be obtained (assuming $\nabla^2\tilde{f}$ is positive definite; see § 4.7) by finding the vector $w = w^*$ that satisfies

$$Z^T\nabla^2\tilde{f}Zw = -Z^T\nabla\tilde{f} \tag{3.3.5}$$

Problem (3.3.1) can then be solved by setting

$$h = Zw^*. \tag{3.3.6}$$

Assuming this solution is different from zero, the direction h is a second-order descent direction for f in the neighbourhood of the point x.

3.4 Dead Points, Dual Estimates and Dropping

Assume we are at some point $\bar{x}$ where the solution to (3.3.5) is zero. Under this condition, $\bar{x}$ is called a dead point for this problem. As in the first-order case, if **A** is full rank, $\nabla f(x)$ can be expressed uniquely as

$$\nabla \tilde{f}(\bar{x}) = \mathbf{A}\bar{u} \tag{3.4.1}$$

What if we are at a point x^k in the neighbourhood of $\bar{x}$ where $I_\epsilon(x^k) \equiv I_\epsilon(\bar{x})$ and the solution to (3.3.5) is "small"? If we define the dual estimate u^k as the least-squares solution to

$$\mathbf{A}u^k = \nabla \tilde{f}(x^k) \tag{3.4.2}$$

then this dual estimate will usually give a reasonable approximation to the dual $\bar{u}$ found in (3.4.1) [see [4], § 2b].

If we are at this dead point $\bar{x}$ and there exists an index $l \in I_\epsilon(\bar{x})$ such that $\| \bar{u}_l \| > \alpha_l$, where $\bar{u}$ is the solution to (3.4.1), then the direction $\bar{h}_l = -\gamma P_l \bar{A}_l \bar{u}_l$ is a descent direction for f at $\bar{x}$ as shown in § 3.2.

If x^k is "sufficiently close" to $\bar{x}$ then $\| u_l^k \| > \alpha_l$ where u^k is the least-squares solution obtained in (3.4.2) when $x = x^k$. Thus there is a neighbourhood of $\bar{x}$ for which

$$\tilde{h}_l = -\gamma P_l \bar{A}_l u_l^k \tag{3.4.3}$$

is a descent direction for f at x^k. (When we use this direction we "drop" $\bar{A}_l$ from **A**.)

3.5 Optimality Conditions and the Linear Refinement

Suppose we are at a point x^* which is a dead point for our function f. In addition, suppose $I_\epsilon(x^*) = \{i_1, \ldots, i_t\}$, $\mathbf{A} = [\bar{A}_{i_1}, \ldots, \bar{A}_{i_t}]$ is full rank and $\| u^*_i \| \leqslant \alpha_i \ \forall i \in I_\epsilon(x^*)$ where $u = u^*$ is a solution to $Au = \nabla \tilde{f}(x^*)$. Then, as long as $I_1(x^*) = \emptyset$, we have (from (3.2.5))

$$f(x + \lambda h) - f(x) = \lambda \left[\sum_{i \in I_\epsilon(x^*)} u^{*T}_i \bar{A}_i^T h + \alpha_i \| \bar{A}_i^T h \| \right] + 0(\lambda^2) \tag{3.5.1}$$

for any choice of direction h. For sufficiently small $\lambda > 0$ this expression must be nonnegative. Since our objective function f is convex, the point x^* must therefore solve our original problem.

However, suppose we are at a point x^k in the neighbourhood of some dead point $\bar{x}$. In addition, assume $\| u_i^k \| \leqslant \alpha_i \ \forall i \in I_\epsilon(x^k)$ where u^k is the least-squares solution to (3.4.2) at this point x^k. As long as x^k was "sufficiently close" to $\bar{x}$ we would then expect $\| \bar{u}_i \| \leqslant \alpha_i \ \forall i \in I_\epsilon(\bar{x})$ where $\bar{u}$ is the solution to (3.4.1). If this were true and $I_1(\bar{x}) = \emptyset$ then $\bar{x}$ would be a solution to our problem! How then should we proceed from the point x^k?

The direction h^k obtained by solving problem (3.3.4) with $x = x^k$ would certainly be a local descent direction for f at the point x^k. It would therefore make sense to take the step $x^k + \lambda_k h^k$ (where λ_k is some computed stepsize) to reduce f. It would also seen appropriate to take some action to force the condition $I_1(x^k + \lambda_k h^k) = \emptyset$. We therefore define the refinement step v^k as the solution to the linear system

$$\begin{bmatrix} \mathbf{A}^T \\ ----- \\ Z^T \end{bmatrix} v^k = \begin{bmatrix} -\overline{r}(x^k + \lambda_k h^k) \\ -------------- \\ 0 \end{bmatrix} \tag{3.5.2}$$

where $\overline{r}$ is the ordered vector of residuals $\overline{r}_i$ (where $\alpha_i \overline{r}_i = r_i$) corresponding to the matrices A_i in $\mathbf{A}$. (The reader should note that the solution to (3.5.2) is the least-squares solution of minimal norm to $\mathbf{A}^T v = -\overline{r}(x^k + \lambda_k h^k)$ since $Z^T v^k = 0$).

With this choice of refinement step we force $I_1(x^k + \lambda_k h^k) = \varnothing$ since

$$\begin{aligned} \overline{r}(x^k + \lambda_k h^k + v^k) &= \overline{r}(x^k + \lambda_k h^k) + \mathbf{A}^T v_k \\ &= 0. \end{aligned}$$

(We try to take $\lambda_k = 1$ in this instance since, for second-order methods, a stepsize of one will asymptotically be optimal; see § 3.7.).

3.6 Degeneracy and Perturbations

Whenever the solution to (3.3.4) is "small" the dual estimate, obtained by solving (3.4.2) in the least-squares sense, becomes important in finding a descent direction or in determining optimality. The uniqueness of this estimate is based on the assumption that the matrix $\mathbf{A} = [\overline{A}_{i_1} \cdots \overline{A}_{i_t}]$, where $I_\epsilon = \{i_1 \cdots i_t\}$, is full rank. This uniqueness is surrendered whenever $\mathbf{A}$ is rank deficient. Fortunately this difficulty can be resolved.

Assume, for the moment, that we are at a point x^k where $\mathbf{A}$ is rank deficient. If we redefine $\mathbf{A} = [\overline{A}_{j_1} \cdots \overline{A}_{j_s}]$, where $J_\epsilon(x^k) = \{j_1 \cdots j_s\} \subset I_\epsilon(x^k)$ and $\mathbf{A}$ forms a basis for the column space of A_i $i \in I_\epsilon(x^k)$, then we can find the least-squares solution to (3.4.2) and obtain the dual estimate u^k which is uniquely defined by this choice of basis. If, after proceeding in this fashion, we find that $\| u_j^k \| \leqslant \alpha_j$ for all $j \in J_\epsilon(x^k)$ then we can safely continue as outlined in § 3.5 (since $A_i^T h = 0 \;\; \forall i \in I_\epsilon(x^k)$). If, however, there exists an index $l \in J_\epsilon(x^k)$ such that $\| u_l^k \| > \alpha_l$, can we then take the direction $h_l = -\gamma P_l \overline{A}_l u_l^k$ as our descent direction under the assumption that (3.2.7) still holds?

The answer to this question is, in general, no. This is because (3.2.7) is based, in part, on the result that, when $\mathbf{A} = [\overline{A}_{i_1} \cdots \overline{A}_{i_t}]$ $\overline{A}_i^T \tilde{h}_l = 0$ for all $i \in I_\epsilon(x^k) - \{l\}$. When $\mathbf{A} = [\overline{A}_{j_1} \cdots \overline{A}_{j_s}]$ we can only guarantee that $\overline{A}_j^T \tilde{h}_l = 0$ for all $j \in J_\epsilon(x^k) - \{l\}$.

Determining an optimal strategy under these circumstances is not a trivial exercise. Here we suggest an approach that is both simple and effective. After taking a refinement step and setting $x^k \leftarrow x^k + v^k$ we randomly perturb the values b_i of all the residuals in the set $I_\epsilon(x^k) \backslash J_\epsilon(x^k)$ so that the gradients $\nabla(\| r_i(x^k) \|)$ are well-defined. This allows these perturbed terms to join the function f when we proceed with our minimization method. Using this approach we either leave this degenerate neighbourhood or identify a solution in this degenerate neighbourhood. In the latter case the point x^k is our solution.

3.7 Minimization Strategy

In order to decrease the objective function at each stage in the minimization process, a decision must be made as to which direction to use. The strategy we suggest here is based, in part, on the analysis presented in [4] and [5].

We consider the following three cases (in all three cases $h = -Z(Z^T \nabla^2 \tilde{f} Z)^{-1} Z^T \nabla \tilde{f}$):

CASE 1: $\| h \| > \beta$

The fact that $\| h \|$ is "large" suggests we are outside the neighbourhood of any dead points. Under this condition we use this direction h to decrease f by setting

$$x \leftarrow x + \lambda h$$

where the stepsize λ is determined via the line search described in § 3.8.

CASE 2: $\| h \| \leqslant \beta$ and $\| u_i \| \leqslant \alpha_i \quad \forall i \in I_\epsilon$

The assumption here is that we are in the neighbourhood of some dead point (which may be optimal). We therefore set

$$\tilde{x} \leftarrow x + h + v$$

where v is the solution to

$$\begin{bmatrix} \mathbf{A}^T \\ \hline Z^T \end{bmatrix} v = \begin{bmatrix} -\bar{r}(x+h) \\ \hline 0 \end{bmatrix}.$$

If $f(\tilde{x}) - f(x) < -\delta_0 \, (\| h \|^2 + r^\infty)$ where δ_0 is some positive constant and $r^\infty = \max\{\| r_i(x) \| \, , \, i \in I_\epsilon(x)\}$) then we accept this as being a "sufficient" decrease and take

$$x \leftarrow \tilde{x};$$

otherwise, we set

$$\epsilon \leftarrow \epsilon/2$$

$$\beta \leftarrow \beta/2$$

and

$$x \leftarrow x + \lambda h$$

where the stepsize λ is determined via the line search described in § 3.8.

CASE 3: $\| h \| \leqslant \beta$ and there exists at least one index $l \in I_\epsilon$ such that $\| u_l \| > \alpha_l$

In this case we define the direction

$$\tilde{h}_l = -\gamma P_l \bar{A}_l u_l.$$

If $g^T \tilde{h}_l < -\delta$, where $\delta = \delta_0 * f(x)/\tau$ and g is defined in § 3.2, then a sufficient decrease along this direction is expected and we set

$$x \leftarrow x + \lambda \tilde{h}_l$$

where the stepsize λ is determined via the line search described in § 3.8. If, on the other hand, $g^T \tilde{h}_l \geqslant -\delta$ for all $l \in I_\epsilon$ with $\| u_l \| > \alpha_l$, we remain at the same point x but set

$$\epsilon \leftarrow \epsilon/2$$

and

$$\beta \leftarrow \beta/2.$$

The parameters β and ϵ are adjusted whenever the step $h + v$ is unsuccessful or the step $\tilde{h}_l$ fails. This, in effect, refines our tests for dead-point neighbourhoods and nondifferentiability.

What follows is a flowchart of our algorithm for solving the continuous multifacility location problem involving Euclidean distances. The performance of this method is affected by the initial choice of the parameters ϵ and β (it is assumed that the chosen values for δ_0 and ϵ_s are "reasonable" and that they therefore have little or no effect on the algorithm's efficiency). The decision as

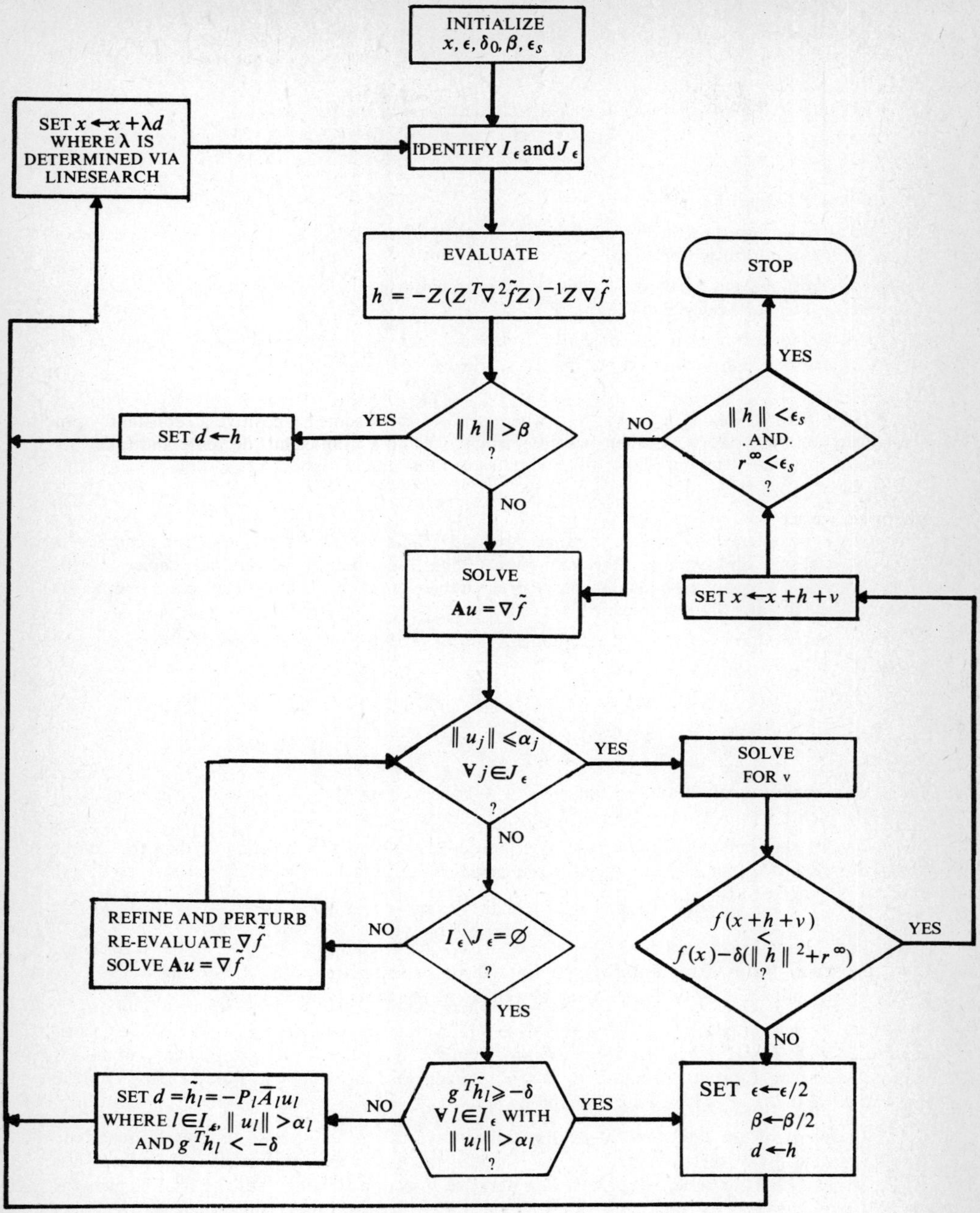

MINIMIZATION FLOWCHART

to what are optimal (or even appropriate) values for the parameters ϵ and β is, by no means, trivial and goes beyond the scope of this paper. It should be noted however that the global convergence properties will be unaffected by this choice. The selection of the starting point is also left up to the user.

The linesearch used in our algorithm is fully described in the next section.

3.8 Line Search Algorithm

In this section we present our method for chosing the steplength λ whenever the line search is invoked in the minimization process.

As in most descent methods our criteria for accepting a steplength is based on convergence requirements (see, for example, [9] and [16]).

In order to ensure that the objective function "decreases sufficiently" with respect to the chosen steplength λ and direction d, we insist that the following condition be met:

$$f(x) - f(x+\lambda d) \geqslant -\mu * \lambda * d^T g(x) \tag{3.8.1}$$

where μ is a preassigned scalar in the range $0 < \mu < 1$. We also ensure that the chosen steplength is large enough by restricting our choice of candidates to those that satisfy the condition:

$$|d^T g(x+\lambda d)| < -\xi * d^T g(x) \tag{3.8.2}$$

where ξ is a preassigned scalar in the range $0 < \xi < 1$. This test also determines the accuracy to which the stepsize approximates the minimum along the line. (The optimal choice of the parameters μ and ξ is not obvious. In our current implementation we have had acceptable results with the values 0.1 and 0.9 respectively.)

Now that we have defined our acceptance criteria lets look at our method of generating trial steplengths.

If we let

$$\Phi'_{+/-}(\cdot) = \lim_{\Delta \to 0^{+/-}} \frac{\Phi(\cdot + \Delta) - \Phi(\cdot)}{\Delta}$$

then λ^* is a minimum of $f(x+\lambda d)$ only if $f'(x+\lambda^* d) = 0$ or λ^* is a derivative discontinuity of $f(x+\lambda d)$ with $f'_-(x+\lambda^* d) \leqslant 0$ and $f'_+(x+\lambda^* d) \geqslant 0$.

It can be shown that if derivative discontinuities exist along the direction d, then they occur at the values $\lambda = \lambda^*_i$ that exactly satisfy the equations

$$r_i(x+\lambda_i d) = r_i(x) + \lambda_i A_i^T d = 0 \quad i \in M \backslash I_\epsilon \tag{3.8.3}$$

(we exclude the set I_ϵ since $r_i(x+\lambda d) = r_i(x) \quad \forall i \in I_\epsilon$). In addition, if $r_i(x+\bar{\lambda} d) = 0$ for some $i \in M \backslash I_\epsilon$ then $r_{i_-}'(x+\bar{\lambda} d) \leqslant 0$ and $r_{i_+}'(x+\bar{\lambda} d) \geqslant 0$.

Consider the set $K = \{ i \in M \backslash I_\epsilon \mid \hat{\lambda}_i > 0 \}$ where the $\hat{\lambda}_i$'s are the least-squares solutions to (3.8.3) (i.e. $\hat{\lambda}_i = -(A_i^T d)^T r_i(x) / \| A_i^T d \|^2 \quad \forall i \in K$). Any number of these $\hat{\lambda}_i$'s may define derivative discontinuities of f along d; moreover, any number may satisfy our acceptance criteria. It therefore seems appropriate to consider, as trial steplengths, the values $\hat{\lambda}_i \; i \in K$. (It should be clear that $\lambda^* \in (0, \hat{\lambda}_{i_{max}}]$ where $i_{max} \in K$ and $\hat{\lambda}_{i_{max}} \geqslant \hat{\lambda}_i \quad \forall i \in K$.)

A flowchart of our linesearch algorithm follows. In this procedure we progress sequentially through a sorted list of the trial steplengths. If one of the trial steplengths satisfies the acceptance criteria, we use it. Otherwise, we perform an iterative bisection starting with the trial steplengths that most closely bound the minimum along the line. This bisection terminates whenever the interval of uncertainty becomes "small" or a bisection point satisfies the acceptance criteria. (The constant ϵ_M which appears in the line search flowchart is defined as the smallest number satisfying

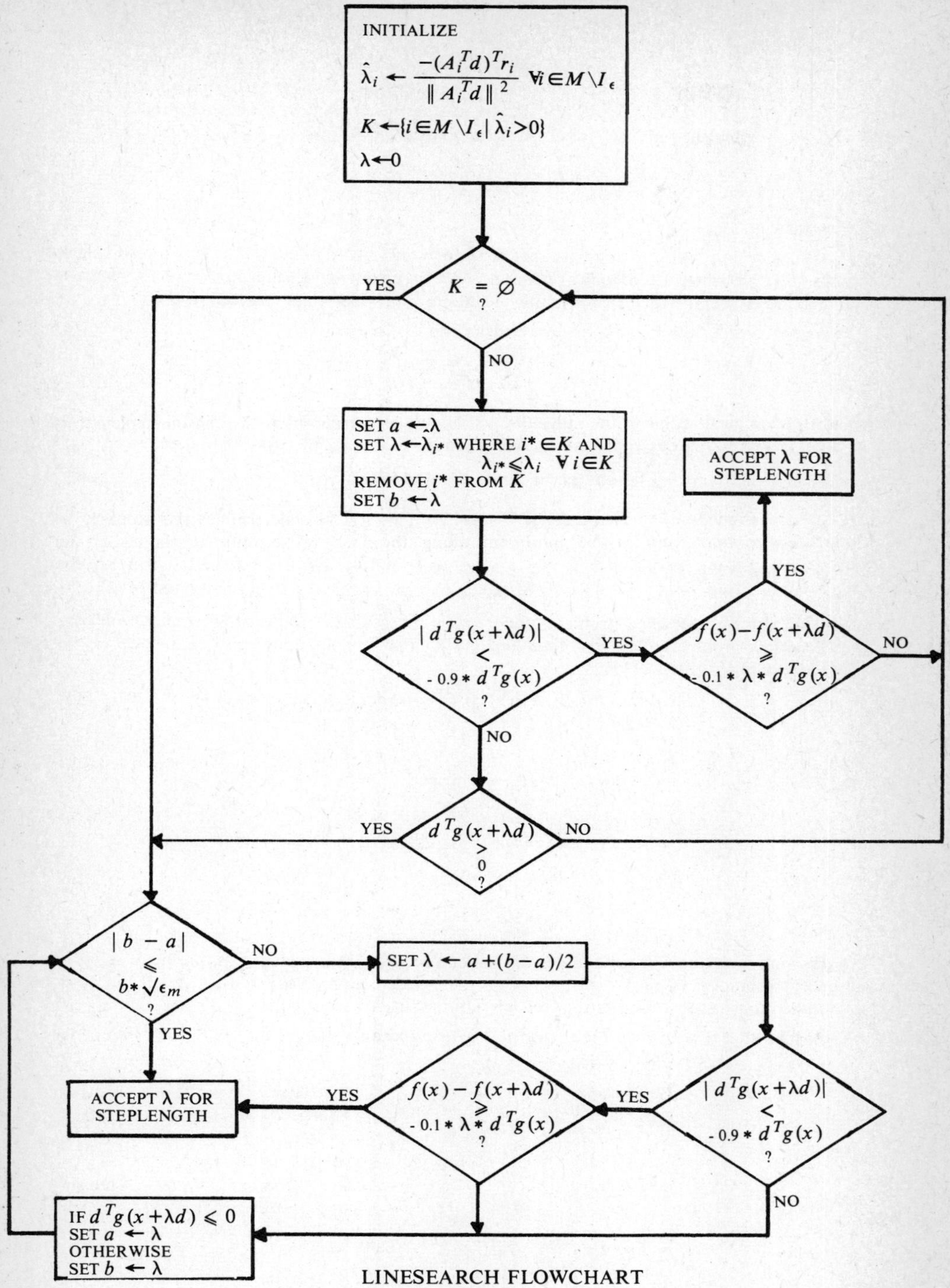

LINESEARCH FLOWCHART

$1 \oplus \epsilon_M > 1$ where $\oplus$ represents floating point addition.)

There are several variants of this linesearch which are currently under investigation by the authors. The most noteworthy involves using, as trial steplengths, only those $\hat{\lambda}_i$'s which "closely" approximate discontinuities along the line (this can be easily determined by computing the values $\| r_i(x + \hat{\lambda}_i d)\|$ to see if they are "small"). If one of these trial steplengths is acceptable, we use it. Otherwise, we bracket the minimum and use an appropriate method to estimate the minimum.

3.9 Extensions

In the foregoing sections of this paper we presented a second-order method for solving the continuous unconstrained multifacility location problem involving Euclidean distances. Here we present some of the details for extending this technique to a wider class of location problems.

First, we consider the linearly constrained location problem:

$$\text{minimize } f(x) = \sum_{i \in M} \| A_i^T x - b_i \| \tag{3.9.1}$$

such that
$$r_i(x) = a_i^T x - b_i \geqslant 0 \quad i \in LI$$
$$r_i(x) = a_i^T x - b_i = 0 \quad i \in LE.$$

If we transform problem (3.9.1) into the unconstrained problem,

$$\text{minimize } F(x, \mu) = \mu f(x) - \sum_{i \in LI} \min[0, r_i(x)] + \sum_{i \in LI} | r_i(x)| \tag{3.9.2}$$

where μ is a positive parameter, then the second-order method we have described can be easily modified to solve problem (3.9.2). Details of the necessary modifications are presented in [2].

As with these constrained problems there are many practical justifications for considering l_p distance location problems (see, for example, [13]). Fortunately, many of the results we have presented are valid for the problem:

$$\text{minimize } f(x) = \sum_{i \in M} \| A_i^T x - b_i \|_p \quad 1 \leqslant p \leqslant \infty \tag{3.9.3}$$

where

$\| \cdot \|_p$ is the general l_p norm, and

$$\| x \|_p = \begin{cases} \left(\sum_{c=1}^{q} | x_c |^p \right)^{1/p} & 1 \leqslant p < \infty \\ \max \{| x_1 |, \ldots, | x_q | \} & p = \infty . \end{cases}$$

Only a few modifications become necessary when we consider solving this problem using our second-order technique when $1 < p < \infty$ or our first-order technique [2] when $p = 1$ or $p = \infty$. These modifications revolve around the following optimality conditions:

The point x^* is a minimum for problem (3.9.3) if and only if

i) $P \nabla \tilde{f}(x^*) = 0$

ii) $I_1(x^*) = \varnothing$

iii) $\| u_i^* \|_{p'} \leqslant \alpha_i \quad \forall i \in I_\epsilon(x^*)$

where

$$p' = \begin{cases} \infty & p = 1, \\ \dfrac{p}{1-p} & 1 < p < \infty, \\ 1 & p = \infty. \end{cases}$$

Thus the minimization strategy for solving problem (3.9.3) involves taking a step in the direction h whenever $\| h \|_2 > \beta$ (CASE 1 § 3.7); attempting the step $h + v$ whenever $\| h \|_2 \leqslant \beta$ and $\| u_i \|_{p'} \leqslant \alpha_i \quad \forall i \in I_\epsilon$ (CASE 2 § 3.7); and attempting a step in the direction $\tilde{h}_l$ whenever $\| h \|_2 \leqslant \beta$ and $\| u_l \|_{p'} > \alpha_l \quad l \in I_\epsilon$ (CASE 3 § 3.7). In this latter instance, $\tilde{h}_l$ is defined as follows:

$$\tilde{h}_l = -\gamma P_l \bar{A}_l \hat{u}_l$$

where, for $1 < p \leqslant \infty$,

$$[\hat{u}_l]_c = \text{sgn}[u_l]_c * | [u_l]_c |^{p'/p} \qquad c = 1, \ldots, q$$

and, for $p = 1$,

$$[\hat{u}_l]_c = \begin{cases} \text{sgn}\,[u_l]_c & c = c^* \\ 0 & \text{otherwise} \end{cases}$$

where $\quad c^*$ is the index to the component of the vector u_l that is largest in magnitude (i.e. $| [u_l]_{c^*} | \geqslant | [u_l]_c | \quad c = 1, \ldots, q$).

As a consequence of this last extension we are also able to solve the continuous multifacility location problem where the distances are combinations of the l_p norms (see [18]). This problem has the form

$$\text{minimize } f(x) = \sum_{k \in K} \sum_{i \in M(k)} \| A_i^T x - b_i \|_{p(k)} \tag{3.9.4}$$

where

$$K = \{k_1, \ldots, k_L\} \qquad L < \infty$$

$$\bigcup_{j=1}^{L} M(k_j) = M$$

$$M(k_j) \cap M(k_{j'}) = \emptyset \qquad j' \neq j \qquad k_j, k_{j'} \in K$$

and

$$1 \leqslant p(k) \leqslant \infty \qquad \forall k \in K$$

Although this particular problem has received very little attention in the literature it is quite obvious that different norms may appear simultaneously in many practical facility location problems.

4. IMPLEMENTATION

4.1 Concepts and Definitions

To explain the implementation of our method we present some basic ideas about graph theory and some additional definitions .

Consider a graph consisting of n vertices having a one-to-one correspondence to the n new facilities (i.e. vertex j corresponds to $NF_j, j = 1, \ldots, n$). Edge $\gamma_{jk}, 1 \leqslant j < k \leqslant n$, is found between vertex j and vertex k if there is an interaction between NF_j and NF_k (i.e. $v_{jk} \neq 0$). A set of edges $\{\gamma_{jk}\}$ form a tree if

a) the edges generate a connected subgraph

b) the edges contain no cycles.

A subgraph is connected if there is a chain joining every pair of distinct vertices in the subgraph. If we consider any sequence of vertices, say $i_1, \ldots, i_l$, then a possible chain would consist of the sequence of edges $\gamma_{i_1 i_2}, \gamma_{i_2 i_3}, \ldots, \gamma_{i_{l-1} i_l}$. The initial vertex of this chain would be i_1 and the terminal vertex would be i_l. A cycle is a chain whose initial vertex and terminal vertex are identical. For example, the sequence of edges $\gamma_{13}, \gamma_{34}, \gamma_{41}$ form a cycle.

Finally let E_j be the $qn \times q$ matrix defined by

$$E_j^T = [\, 0_{j-1} \quad I_q \quad 0_{n-j} \,]$$

where

0_k is a zero matrix of dimension $q \times qk$

and

I_q is a $q \times q$ identity matrix,

and let the sets $J(*)$, $-\lfloor n/2 \rfloor \leqslant * \leqslant \lfloor n/2 \rfloor$, and K^* be defined as

$$J(*) = \{j \mid \text{TREE}(j) = *\} = \{\beta_1, \dots, \beta_{J_*}\}$$

and

$$K^* = \{k \mid J(k) \neq \emptyset\}$$

where

$$\beta_i = \beta_i(*) \qquad i = 1, \dots, J_*$$

$$\beta_1 < \beta_2 < \cdots < \beta_{J_*}$$

$$J_* = |\, J(*) \,|$$

and where the vector TREE is defined in § 4.2.

4.2 Active Trees and Masking

Each term $\| r_i \|$ $i \in M$ in our objective function represents either an interaction between two new facilities ($i \leqslant \eta$) or an interaction between a new facility and some existing facility ($\eta < i \leqslant \tau$). In the former instance the interaction between the two *NF*'s is represented by the edge in our graph joining the two *NF* vertices. Those two *NF*'s are included as vertices in an *active* tree under the following conditions:

a) $\| r_i \| \leqslant \epsilon$

b) the inclusion of the edge does not form a cycle

c) both *NF*'s are not *masked* (this term is defined below).

When the objective function term involves the interaction between a *NF* and some *EF* we *mask* the *NF* vertex down 1 under the following conditions:

a) $\| r_i \| \leqslant \epsilon$

b) the vertex is not already masked.

In both instances the first condition implies that the corresponding objective function term is nondifferentiable (or near-nondifferentiable) whereas the remaining conditions detect degeneracies.

Now suppose that at each stage in our minimization process we have the n-vector TREE whose i-th element is set according to the following rules:

a) TREE(i) $\leftarrow$ 0 — if and only if NF_i is not a vertex in any active tree and is not masked by any *EF*.

b) TREE(i) $\leftarrow k$ — if and only if NF_i is a vertex in the k-th active tree and no *NF* in that tree is masked by any *EF*.

c) $\text{TREE}(i) \leftarrow -k$ if and only if NF_i is a vertex in the k-th active tree and some NF in that tree is masked by some EF.

d) $\text{TREE}(i) \leftarrow -\lfloor n/2 \rfloor -1$ if and only if NF_i is not a vertex in any active tree but is masked by some EF.

(Since there can be no more than $\lfloor n/2 \rfloor$ distinct trees in a graph with n vertices we satisfy $0 \leqslant k \leqslant \lfloor n/2 \rfloor$ in the foregoing definitions.)

As we shall see in subsequent sections, once we have the n-vector TREE we can very easily:

1) Identify I_ϵ,
2) Form a basis **A** for A_i, $i \in I_\epsilon$,
3) Construct Z.

The method used for constructing the n-vector TREE is a simple adaptation of the classical Spanning Tree Algorithm (see [14]). What follows is a flowchart for this construction. In this flowchart (and in subsequent subsections) i^*,j^* and k^* are defined as they were in § 2.

4.3 Identifying I_ϵ

For each term $\| r_i \|$, $i = 1, \ldots, \tau$ in our objective function we have defined the values j^* and k^* (for $i \leqslant \eta$) and j^* and i^* (for $\eta < i \leqslant \tau$). We add the index i to the set I_ϵ under either of the following two conditions:

$$\text{If} \quad i \leqslant \eta \quad \text{and} \quad \text{TREE}(j^*) = \text{TREE}(k^*) \neq 0 \quad \text{or} \quad \text{TREE}(j^*) < 0 \text{ and } \text{TREE}(k^*) < 0 . \tag{4.3.1}$$

$$\text{If} \quad \eta < i \leqslant \tau \quad \text{and} \quad \text{TREE}(j^*) < 0 . \tag{4.3.2}$$

A brief examination of the TREE VECTOR FLOWCHART should convince the reader that all terms $\| r_i \|$, $i = 1, \ldots, \tau$ satisfying the inequality $\| r_i \| \leqslant \epsilon$ will also satisfy (4.3.1) or (4.3.2). However, there can be indices $i \in I_\epsilon$ that do not satisfy this inequality. Consider, for instance, the following situation:

Example 1: $n = 2$, $m = 1$, $q = 2$, $p = 2$. $v_{12} = w_{11} = w_{21} = 1$. $x_1^T = \{\epsilon, 0\}$, $x_2^T = \{0, \epsilon\}$, $p_1^T = \{0,0\}$. We have $\| r_1 \| = \sqrt{2} * \epsilon$ and $\| r_2 \| = \| r_3 \| = \epsilon$. We also have TREE(1) = TREE(2) = -2. Now, for $i = 1$, $j^* = 1$, $k^* = 2$ and $\text{TREE}(j^*) = \text{TREE}(k^*) \neq 0$. Therefore, according to (4.3.1), the index $i = 1$ is added to I_ϵ even though $\| r_1 \| > \epsilon$.

The reasoning for this apparent inconsistency is simple. Consider Example 1 again. Since $\| r_2 \| \leqslant \epsilon$ and $\| r_3 \| \leqslant \epsilon$ index 2 and 3 belong in I_ϵ. The fact that $\overline{A}_i^T d = 0 \ \forall i \in I_\epsilon$ guarantees that $\overline{A}_1^T d = 0$ since $\overline{A}_1 = \overline{A}_2 - \overline{A}_3$. Thus index 1 can be added to I_ϵ without increasing the rank or complicating the construction of **A** (see § 4.4). In addition, whenever a facility is in the "neighbourhood" of another facility we have a term that is potentially nondifferentiable and whose index may therefore belong in I_ϵ. Since x_1 and x_2 are both in ϵ−neighbourhoods of P_1 (i.e. $\| r_2 \| = \| r_3 \| = \epsilon$) indices 2 and 3 are added to I_ϵ. In making this decision we, in effect, "pretend" that x_1 and x_2 coincide with P_1. Under this pretension, x_1 must coincide with x_2 (making $\| r_1 \| = 0$). We may therefore include index 1 in I_ϵ.

It can be shown that $\forall i \in I_\epsilon$ either:

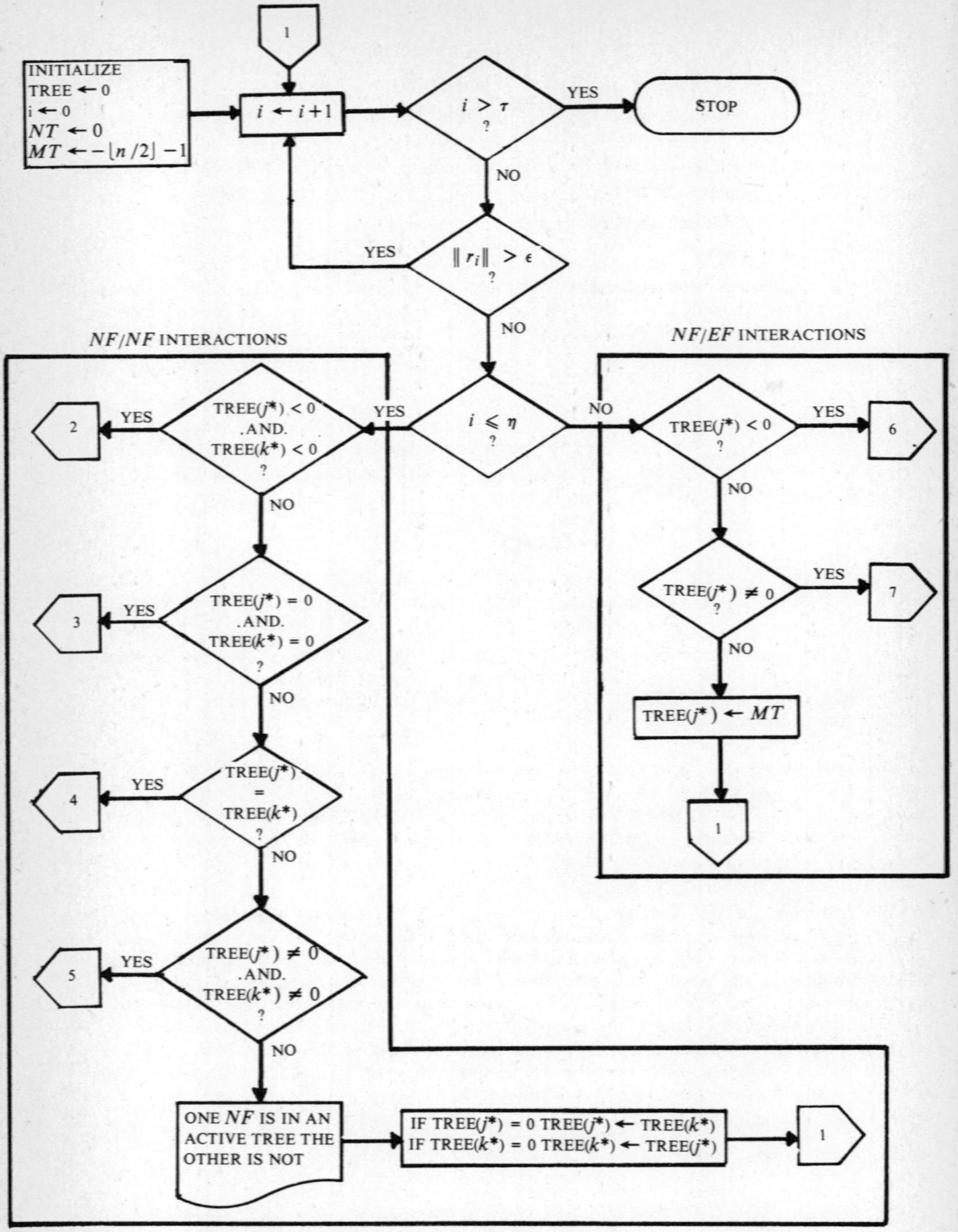

TREE VECTOR FLOWCHART

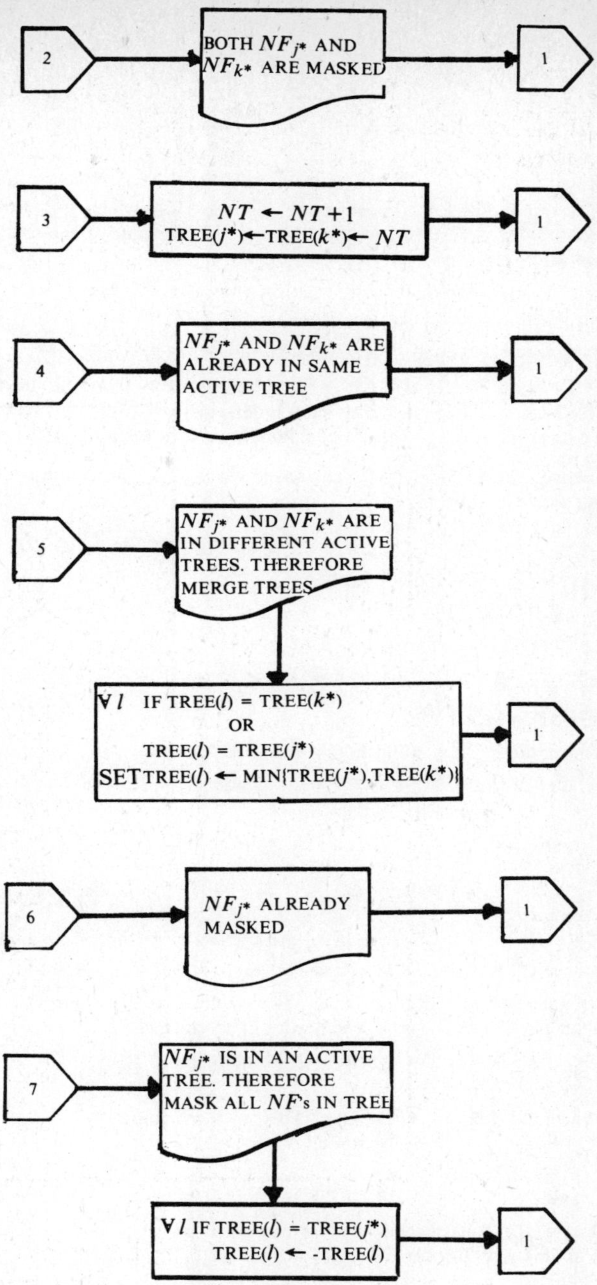
2
BOTH NF_{j^*} AND NF_{k^*} ARE MASKED
1
3
$NT \leftarrow NT+1$
TREE(j^*)←TREE(k^*)← NT
1
4
NF_{j^*} AND NF_{k^*} ARE ALREADY IN SAME ACTIVE TREE
1
5
NF_{j^*} AND NF_{k^*} ARE IN DIFFERENT ACTIVE TREES. THEREFORE MERGE TREES
∀ l IF TREE(l) = TREE(k^*)
OR
TREE(l) = TREE(j^*)
SET TREE(l) ← MIN{TREE(j^*),TREE(k^*)}
1
6
NF_{j^*} ALREADY MASKED
1
7
NF_{j^*} IS IN AN ACTIVE TREE. THEREFORE MASK ALL NF's IN TREE
∀ l IF TREE(l) = TREE(j^*)
TREE(l) ← -TREE(l)
1

TREE VECTOR FLOWCHART (continued)

a) $\| r_i \| \leqslant \epsilon$

or b) there exists vectors y_i such that $\overline{A}_i = \sum_{j \in I_\epsilon^*} \overline{A}_j y_j$ where $I_\epsilon^* = \{i \in I_\epsilon | \; \| r_i \| \leqslant \epsilon\}$.

4.4 Constructing A

What follows is our method for constructing the matrix **A** which is a basis for the column space of the matrices $\overline{A}_i \;\; \forall i \in I_\epsilon$. This method takes full advantage of the structure of this problem and its graphical interpretation.

For $j = 1, \ldots, n$, the j-th step of the construction process is as follows:

a) If TREE(j) = 0 no contribution is made to **A**.

b) If TREE(j) < 0 augment **A** with the matrix E_j.

c) If TREE(j) = l , $1 \leqslant l \leqslant \lfloor n/2 \rfloor$, then let $k > j$ be the next index satisfying TREE(k) = TREE(j). If no such index exists then proceed to the next step in the construction process, otherwise augment **A** with the matrix $[E_j - E_k]$.

We can now prove that **A** is a basis for the space spanned by the columns of $\overline{A}_i \;\; \forall i \in I_\epsilon$ (i.e. $\forall i \in I_\epsilon \;\; \overline{A}_i \in C(\mathbf{A})$ and **A** is full rank.)

Proof that $\overline{A}_i \in C(\mathbf{A}) \;\; \forall i \in I_\epsilon$

For $i \in I_\epsilon^* \;\; \| r_i \| \leqslant \epsilon$ and either

1) : $i \leqslant \eta$ and TREE(j^*) = TREE(k^*) > 0 .

or 2) : $i \leqslant \eta$ and TREE(j^*) < 0 and TREE(k^*) < 0 .

or 3) : $i > \eta$ and TREE(j^*) < 0 .

For case 1), assume that TREE(j^*) = TREE(k^*) = l and let $\{j_1, j_2, \cdots, j_L\} = \{ j \mid \text{TREE}(j) = l, j^* \leqslant j \leqslant k^* \}$ where $j_1 < j_2 < \cdots < j_L$.

Thus

$$\begin{aligned} \overline{A}_i &= E_{j_1} - E_{j_L} \\ &= [E_{j_1} - E_{j_2}] + [E_{j_2} - E_{j_3}] + \cdots + [E_{j_{L-1}} - E_{j_L}]. \end{aligned}$$

But $[E_{j_1} - E_{j_2}], [E_{j_2} - E_{j_3}], \ldots, [E_{j_{L-1}} - E_{j_L}]$ are the matrices that were augmented to **A** in its construction process when $j = j_1, \ldots, j_{L-1}$ respectively. Therefore $\overline{A}_i \in C(\mathbf{A})$.

In case 2), $\overline{A}_i = E_{j^*} - E_{k^*}$. But E_{j^*} and E_{k^*} are the matrices augmented to **A** in its construction process when $j = j^*$ and $j = k^*$ respectively. Therefore $\overline{A}_i \in C(\mathbf{A})$.

For case 3), $\overline{A}_i = E_{j^*}$. But E_{j^*} is the matrix augmented to **A** in its construction process when $j = j^*$. Thus $\overline{A}_i \in C(\mathbf{A})$.

We have shown that $\forall i \in I_\epsilon^* \; \overline{A}_i \in C(\mathbf{A})$. This, in turn, guarantees that $\overline{A}_i \in C(\mathbf{A}) \;\; \forall i \in I_\epsilon$ since, for $i \in I_\epsilon \backslash I_\epsilon^* \;\; \overline{A}_i = \sum_{j \in I_\epsilon^*} \overline{A}_j y_j$ for some vectors y_j.

Proof that A is full rank

At the j-th step in the construction process there are, at most, $q(j-1)$ columns in **A**. Any augmentation to **A** at this j-th step involves the matrix E_j or the matrix $[E_j - E_k]$ both of which have zeros in rows 1 through $q(j-1)$ and a $q \times q$ identity matrix in the next q rows. Thus **A** is full rank (and lower-trapezoidal) and rank(**A**) = $q * (n_1 + n_2)$ where

$n_1 \triangleq$ number of *NF*'s that are masked but not in active trees, and

$n_2 \triangleq$ number of *NF*'s in active trees minus the number of unmasked active trees.

4.5 Constructing Z

The matrix Z, defined by (3.3.2) and (3.3.3), is an orthonormal basis for the null space of the matrix $\mathbf{A}$ (i.e. $\mathbf{A}^T Z = 0$ and $Z^T Z = I_{q(n-t)}$ where rank($\mathbf{A}$) $= qt = q(n_1 + n_2)$). In [2] it is suggested that Z be computed using the QR factorization of $\mathbf{A}$. However, the structure of the problem allows a much more efficient method for constructing Z.

We construct the matrix Z, using the information in the vector TREE, as follows:

Step 1. If there are no zero entries in the vector TREE proceed to Step 2; otherwise, $\forall k \in \{j \mid \text{TREE}(j) = 0\}$ augment Z with the matrix E_k.

Step 2. If there are no strictly positive entries in the vector TREE, the construction process is complete; otherwise, for $l = 1, \ldots, \lfloor n/2 \rfloor$ if $J(l) \neq \emptyset$ augment Z with the matrix $\frac{1}{\sqrt{J_l}} \sum_{j \in J(l)} E_j$.

With this construction process in mind we can prove that the matrix Z is, in fact, a basis for the null space of $\mathbf{A}$. (The proof that Z is orthonormal is trivial.)

Proof that $A^T Z = 0$ and rank$(Z) = q(n-t)$

The matrix Z has only zero entries in rows $q(j-1)+1$ through qj when TREE(j) < 0. Therefore $E_j^T Z = 0$ for all indices j satisfying TREE(j) < 0. Similarily, there are only q columns in Z with nonzero entries in rows $q(j-1)+1$ through qj or $q(k-1)+1$ through qk when TREE(j) $=$ TREE(k) $= l > 0$. These nonzero entries are all in the matrix $\frac{1}{\sqrt{J_l}} \sum_{j \in J(l)} E_j$ that was augmented to Z in step 2 of its construction process. Therefore $(E_j - E_k)^T Z = 0$ for all indices j and k satisfying TREE(j) $=$ TREE(k) $= l > 0$. Since $\mathbf{A}$ is composed only of submatrices E_j, where TREE(j) < 0 , and $(E_j - E_k)$, where TREE(j) $=$ TREE(k) $= l > 0$, we have $\mathbf{A}^T Z = 0$.

Each matrix augmenting Z in the construction process has rank q and no two such matrices have nonzero entries in any of the same row positions, therefore

$$\text{rank}(Z) = q * \text{number of matrices augmented to } Z$$

$$= q * (n_3 + n_4)$$

where

$n_3 \triangleq$ number of *NF*s that are unmasked and do not belong to any active trees, and

$n_4 \triangleq$ number of unmasked active trees.

But

$$(n_3 + n_4) = n - (n_1 + n_2)$$

$$= n - t.$$

Therefore rank$(Z) = q(n-t)$.

4.6 The Direction $\tilde{h}_l$

Our minimization algorithm sometimes requires an iteration using the direction

$$\tilde{h}_l = -\gamma P_l \bar{A}_l u_l. \tag{4.6.1}$$

As we shall see below, the vector TREE is easily modified to indicate index l no longer belongs in the set I_ϵ (i.e. the columns of $\bar{A}_l$ are to be dropped from the matrix **A**). Once this is accomplished the matrix Z can be reconstructed. Since Z then forms an orthogonal basis for the space $S_l^{\perp}$, the projection matrix P_l can be computed as

$$P_l = Z Z^t \tag{4.6.2}$$

and the direction $\tilde{h}_l$ can be formed.

In section 4.4. we learned that the matrix $\bar{A}_l$ was constructed from either the matrix E_j where $1 \leqslant j \leqslant n$ or the matrix $(E_j - E_k)$, where $1 \leqslant j < k \leqslant n$. With this thought in mind the vector TREE is modified as follows:

1) If $\bar{A}_l$ is constructed from E_j then set TREE(j) $\leftarrow 0$.
2) If $\bar{A}_l$ is constructed from $[E_j - E_k]$ then let $l^* \in \{l \mid 1 \leqslant l \leqslant \lfloor n/2 \rfloor, l \notin K^*, -l \notin K^*\}$ and set TREE(i) $\leftarrow l^*$ for all $i \geqslant k$ with TREE(i) = TREE(j).

4.7 The Second-Order Direction

The direction h which is obtained by finding the solution $w = w^*$ to

$$Z^T \nabla^2 \tilde{f} Z w = -Z^T \nabla \tilde{f} \tag{4.7.1}$$

and then setting

$$h = Z w^* \tag{4.7.2}$$

is called the projected Newton direction (ie h is the Newton step to the minimum of problem (3.3.1)). The matrix $Z^T \nabla^2 \tilde{f} Z$ and the vector $Z^T \nabla \tilde{f}$ are respectively called the projected Hessian and the projected gradient.

Since f is convex, $Z^T \nabla^2 \tilde{f} Z$ is positive semi-definite. When $Z^T \nabla^2 \tilde{f} Z$ is not positive definite or when it is positive definite but very ill-conditioned, we cannot (stably) solve (4.7.1). We can, however, apply a numerically stable modified Cholesky factorization to $Z^T \nabla^2 \tilde{f} Z$ to obtain the system

$$L D L^T w = -Z^T \nabla \tilde{f} \tag{4.7.3}$$

where $LDL^T = Z^T \nabla^2 \tilde{f} Z + E$, L is a lower-triangular matrix, D is a diagonal matrix and E is a diagonal matrix that is zero when $Z^T \nabla^2 \tilde{f} Z$ is "sufficiently" positive definite. The solution $w = w^*$ to (4.7.3) can then be computed, using a forward and backward substitution, and used to obtain the modified projected Newton direction $h = Zw^*$. (The reader is referred to [10] for a complete description of the modified Cholesky factorization.)

It should be noted that at each stage in the minimization process we use the true projected Hessian and not some approximation to it (it is possible to use some Quasi-Newton approach). This choice is justifiable since the construction of Z and the computation of $\nabla^2 \tilde{f}$ are relatively inexpensive.

4.8 The Refinement Step

In section 3.5 we showed that the refinement step v could be obtained by solving the linear system

$$\begin{bmatrix} \mathbf{A}^T \\ \hline Z^T \end{bmatrix} v = \begin{bmatrix} -\bar{r}(x+h) \\ \hline 0 \end{bmatrix}. \tag{4.8.1}$$

As we shall soon see, solving this system is a trivial process as a result of its structure.

If we let $v^T = \{v_1^T \cdots v_n^T\}$ and $\bar{r}(x+h)^T = \{\bar{r}_1^T \cdots \bar{r}_t^T\}$ where v_j^T and $\bar{r}_i^T$ are $1 \times q$ vectors, then we can solve for v as follows:

Step 1. If $J(l) = \varnothing$ for $l = -1, \ldots, -\lfloor n/2 \rfloor$ then go to Step 2; otherwise, $\forall j \in J(l)$, $l = -1, \ldots, -\lfloor n/2 \rfloor$, the matrix E_j augmented $\mathbf{A}$ in its construction (see § 4.4). If E_j was the ξ_j^{th} matrix to augment $\mathbf{A}$ then, from (4.8.1), we have $E_j^T v = v_j = -\bar{r}_{\xi_j}$.

Step 2. If $J(0) = \varnothing$ go to step 3; otherwise, for all $j \in J(0)$ the matrix E_j augmented Z in its construction (see § 4.5). Then, from (4.8.1), we have

$$E_j^T v = v_j = 0.$$

Step 3. For $l = 1, \ldots, \lfloor n/2 \rfloor$, if $J(l) \neq \varnothing$ then the matrix $\frac{1}{\sqrt{J_l}} \sum_{j \in J(l)} E_j$ augmented Z in its construction (see § 4.5) and the matrices $[E_{\beta_1} - E_{\beta_2}], \ldots, [E_{\beta_{J_l-1}} - E_{\beta_{J_l}}]$ augmented $\mathbf{A}$ in its construction (see § 4.4). If, for $i = 1, \ldots, J_l - 1$, the matrix $[E_{\beta_i} - E_{\beta_{i+1}}]$ was the $\xi_{\beta_i}^{\text{th}}$ matrix to augment $\mathbf{A}$ then, from 4.8.1, we have

$$\begin{bmatrix} E_{\beta_1}{}^T - E_{\beta_2}{}^T \\ \vdots \\ E_{\beta_{J_l-1}}{}^T - E_{\beta_{J_l}}{}^T \\ E_{\beta_1}{}^T + \cdots + E_{\beta_{J_l}}{}^T \end{bmatrix} v = \begin{bmatrix} -\bar{r}_{\xi_{\beta_1}} \\ \vdots \\ -\bar{r}_{\xi_{\beta_{J_l-1}}} \\ 0 \end{bmatrix}$$

or, equivalently,

$$v_{\beta_{J_l}} = \left[\bar{r}_{\xi_{\beta_1}} + 2\bar{r}_{\xi_{\beta_2}} + \cdots + (J_l - 1)\bar{r}_{\xi_{\beta_{J_l-1}}}\right] / J_l$$

and

$$v_{\beta_i} = v_{\beta_{i+1}} - \bar{r}_{\xi_{\beta_i}} \qquad i = J_l - 1, \ldots, 1.$$

4.9 The Dual Estimates

In our minimization algorithm we sometimes compute the least-squares solution to the system

$$\mathbf{A}u = \nabla \tilde{f}. \tag{4.9.1}$$

If we factor $\mathbf{A}$ into the product

$$Q^T \mathbf{A} = P \begin{bmatrix} R \\ 0 \end{bmatrix}, \tag{4.9.2}$$

where Q^T is an orthonormal matrix, P is a permutation matrix and R is an upper-triangular matrix, then this least-squares solution is efficiently obtained by solving the system

$$P\begin{bmatrix} R \\ 0 \end{bmatrix} u = Q^T \nabla \tilde{f} \,. \tag{4.9.3}$$

To describe the factorization given by (4.9.2), let the Givens reflection matrices $\overline{G}_i$, $i = 1, 2, \cdots$ be defined as

$$\overline{G}_i = \begin{bmatrix} c_i I_q & s_i I_q \\ s_i I_q & -c_i I_q \end{bmatrix}$$

where

$$c_0 = 0, \qquad s_0 = 1$$

$$c_i = \frac{1}{\sqrt{(1+s_{i-1}^2)}}, \quad s_i = \frac{-s_{i-1}}{\sqrt{(1+s_{i-1}^2)}} \quad i = 1, 2, \ldots,$$

and let the matrix $G_i(j,k)$ be the $qn \times qn$ matrix obtained by imbedding $\overline{G}_i$ in the qn-dimensional identity matrix as follows:

$$G_i(j,k) = \begin{array}{c} \\ 1 \\ \\ \\ j \\ \\ \\ \\ \\ k \\ \\ \\ n \end{array} \overset{\begin{array}{cccccccccc} 1 & & j & & & & k & & & n \end{array}}{\begin{bmatrix} I_q & & & & & & & & \\ & \ddots & & & & & & & \\ & & I_q & & & & & & \\ & & & c_i I_q & & & s_i I_q & & \\ & & & & I_q & & & & \\ & & & & & \ddots & & & \\ & & & & & & I_q & & \\ & & & s_i I_q & & & -c_i I_q & & \\ & & & & & & & I_q & \\ & & & & & & & & \ddots \\ & & & & & & & & & I_q \end{bmatrix}}$$

We then take

$$Q^T = \prod_{\substack{k \in K^* \\ k > 0}} \prod_{i=1}^{J_k - 1} G_{J_k - 1}(\beta_i, \beta_{J_k})$$

where K^*, J_k and β_i are as defined in § 4.1.

The effect of premultiplying **A** by Q^T is equivalent to having augmented **A** with the matrices

$$\left[\left(c_{J_k-1} - s_{J_k-1} s_{J_k-2}\right) E_{\beta_1} - c_{J_k-2} E_{\beta_2}\right], \ldots, \left[\left(c_1 - s_1 s_0\right) E_{\beta_{J_k-1}} - c_0 E_{\beta_{J_k}}\right] \tag{4.9.4}$$

instead of the matrices

$$\left[E_{\beta_1} - E_{\beta_2}\right], \ldots, \left[E_{\beta_{J_k-1}} - E_{\beta_{J_k}}\right] \tag{4.9.5}$$

when $j = \beta_1, \beta_2, \ldots, \beta_{J_k - 1}$ in **A**'s construction (see § 4.4). Thus we form $Q^T\mathbf{A}$ or more appropriately, $P\begin{bmatrix} R \\ 0 \end{bmatrix}$ by transforming **A** using the relationships given by (4.9.4) and (4.9.5) for all $k \in K^*$, $k > 0$.

If we compare (4.9.4) and (4.9.5) it becomes clear that there is no fill-in whatsoever when **A** is transformed in the described manner. However, since $c_0 E_{\beta_{J_k}} = 0 \; k \in K^* \; k > 0$, each augmentation of the matrices given by (4.9.4) results in q zero rows replacing q nonzero rows of **A** (These q nonzero rows of **A** contained, for their nonzero entries, the matrix $-I_q$ which resulted from augmenting **A** with the matrix $\left[E_{\beta_{J_k - 1}} - E_{\beta_{J_k}}\right]$.) It is the introduction of these zero rows that allows us to form the factorization given by (4.9.2).

As a result of the manner in which **A** is formed (§ 4.4) and transformed (by premultiplication by Q^T) the upper-triangular matrix **R** is obtained by simply disregarding the zero rows of $Q^T\mathbf{A}$ (ie the permutation matrix P does not re-order the rows of **R**). Therefore, once $Q^T\nabla f$ is computed we can solve for the dual estimate u in (4.9.3) by simple forward-substitution.

If we let $\nabla\tilde{f} = [f_1^T \cdots f_n^T]^T$ and $Q^T\nabla\tilde{f} = [\bar{f}_1^T \cdots \bar{f}_n^T]^T$, where f_i and $\bar{f}_i$ are $q \times 1$ vectors, then $Q^T\nabla f$ is obtained by performing the following algorithm:

SET $\bar{f}_i \leftarrow f_i \quad i = 1, \ldots, n.$
DO for all $k \in K^*$, $k > 0$
  DO for $i = 1, \ldots, J_K - 1$
    TEMP $\leftarrow c_i \bar{f}_{\beta_{J_k - i}} + s_i \bar{f}_{\beta_{J_k}}$
    $\bar{f}_{\beta_{J_k}} \leftarrow s_i \bar{f}_{\beta_{J_k - i}} - c_i \bar{f}_{\beta_{J_k}}$
    $\bar{f}_{\beta_{J_k - i}} \leftarrow$ TEMP

5. PRELIMINARY NUMERICAL RESULTS

In this section we provide a cursory comparison between the performance of the projected Newton method (PNM) described in this paper, the hyperboloid approximation procedure (HAP [6]) and a projected steepest descent method (PSDM [2]). These three algorithms were implemented in FORTRAN on a Honeywell 66/60 using single precision arithmetic.

Six small problems were run as a basis for this comparison. The first three problems are given in [8] (as exercises #5.23, #5.6 and #5.7 respectively), the fourth is reported in [6] and the last two appear in [1].

The results of these test runs are summarized in Table 1. Except for the last row, the figures in the table refer to the number of iterations required to reach the solution.

In the last row an estimate of the total number of addition operations (in units of one thousand) required in solving the six problems is given (approximately the same number of multiplications would be required). The reader should note that the number of addition operations quoted in [2], for the projected steepest descent method, is greater than the number quoted here. This is because the structure of the problem was not taken into account when this method was originally implemented and tested.

In all problems, except #5, the projected Newton method outperformed the other methods in terms of both the number of iterations and the number of addition operations. The performance on problem #5 could be improved by an alternate choice of the free parameters (for all six problems the free parameters for PNM were set as follows: $\epsilon = 10^{-1}, \beta = 10^{-2}, \delta_0 = 10^{-5}, \epsilon_s = 10^{-8}$ and $\epsilon_M = 7.45 \times 10^{-9}$).

#	HAP	PSDM	PNM
1	1661	64	17
2	647	17	6
3	87	8	4
4	45	17	12
5	142	26	29
6	242	18	6
TOTAL	2824	150	74
+OPs/1000	387	49	59

Table 1 Comparative Test Results

A much more thorough investigation into the performance of this second-order method is currently under way and is intended for future publication.

6. CONCLUDING REMARKS

Our objective has been to provide a unified and numerically stable approach for solving facility location problems. To achieve this goal we have presented a second-order method, involving projected Newton steps, that can be applied to a wide class of location problems. For efficiency, the method has been designed to exploit the sparsity and structure that are inherent in all these problems regardless of their scale. In addition, the degeneracies that occur quite frequently in multifacility location problems are easily resolved using the proposed method.

ACKNOWLEDGEMENTS

The authors would like to thank Mary Wang, Brian Finch and Marney Heatley for their help in typesetting this paper using Troff and the Photon Econosetter.

REFERENCES

[1] Calamai, P.H., and Charalambous, C., "Solving Multifacility Location Problems Involving Euclidean Distances", Naval Res. Log. Quart., 27, 609-620, (1980).

[2] Calamai, P.H., and Conn, A.R., "A Stable Algorithm for Solving the Multifacility Location Problem Involving Euclidean Distances", SIAM J. Sci. Stat. Comput., 1, 512-525, (1980).

[3] Chatelon, J.A., Hearn, D.W., and Lowe, T.J., "A Subgradient Algorithm for Certain Minimax and Minisum Problems", Math Prog., 15, 130-145, (1978).

[4] Coleman, T.F., and Conn, A.R., "Nonlinear Programming Via an Exact Penalty Function: Global Analysis", Math. Prog., (to appear).

[5] Coleman, T.F., and Conn, A.R., "Nonlinear Programming Via an Exact Penalty Function: Asymptotic Analysis", Math. Prog., (to appear).

[6] Eyster, J.W., White, J.A., and Wierwille, W.W., "On Solving Multifacility Location Problems Using a Hyperboloid Approximation Procedure", AIIE Trans., 5, 1-6, (1973).

[7] Francis, R.L., and Goldstein, J.M., "Location Theory: A Selective Bibliography", Oper. Res., 22, 400-410, (1974).

[8] Francis, R.L., and White, J.A., "Facility Layout and Location: An Analytic Approach", Prentice-Hall, New Jersey, (1974).

[9] Gill, P.E., and Murray, W., "Safeguarded Steplength Algorithms for Optimization Using Descent Methods", Report NAC 37, National Physical Laboratory, England, (1974).

[10] Gill, P.E., and Murray, W., "Newton-type Methods for Unconstrained and Linearly Constrained Optimization", Math. Prog., 7, 311-350, (1974).

[11] Lea, A.C., "Location-Allocation Systems: An Annotated Bibliography", Discussion Paper No. 13, Univ. of Toronto, Department of Geography, Canada, (1973).

[12] Love, R.F., "Locating Facilities in Three-dimensional Space by Convex Programming", Naval Res. Log. Quart., 16., 503-516, (1969).

[13] Love, R.F., and Morris, J.G., "Modelling Inter-city Road Distances by Mathematical Functions", Opnl. Res. Quart., 23, 61-71, (1972).

[14] Minieka, E., "Optimization Algorithms for Networks and Graphs", Industrial Engineering Series: Volume 1, Marcel Dekker Inc., (1978).

[15] Morris, J.G., "A Linear Programming Solution to the Generalized Rectangular Distance Weber Problem", Naval Res. Log. Quart., 22, 155-164, (1975).

[16] Murray, W., and Overton, M.L., "Steplength Algorithms for Minimizing a Class of Non-Differentiable Functions", Computing, 23, 309-331, (1979).

[17] Overton, M.L., "A Quadratically Convergent Method for Minimizing a Sum of Euclidean Norms", Tech. Report #030, Dept. of Comp. Sci., Courant Inst. of Math. Sci., (1981).

[18] Planchart, A., and Hurter, A.P., "An Efficient Algorithm for the Solution of the Weber Problem With Mixed Norms", SIAM J. Control, 13, 650-665, (1975).

[19] Vergin, R.C., and Rogers, J.D., "An Algorithm and Computational Procedure for Locating Economic Facilities", Management Sci., 13, B240-B254, (1967).

[20] Wesolowsky, G.O., and Love, R.F., "A Nonlinear Appoximation Method for Solving a Generalized Rectangular Distance Weber Problem", Management Sci., 18, 656-663, (1972).

[21] Wesolowsky, G.O., and Love, R.F., "The Optimal Location of New Facilities Using Rectangular Distances", Oper. Res., 19, 124-130, (1971).

DATA SMOOTHING BY DIVIDED DIFFERENCES

M.P. Cullinan and M.J.D. Powell

1. Introduction

Let $\{\phi_i; i = 1,2,\ldots,n\}$ be measurements of the function values $\{f(x_i); i = 1,2,\ldots,n\}$, where f is a real function of one variable, and where the abscissae $\{x_i; i = 1,2,\ldots n\}$ are distinct and in ascending order. An excellent way of determining whether the measurements are smooth is to form a divided difference table of the data (see Hildebrand, 1956, for instance). A single random error tends to cause k sign changes in the divided differences of order k that are affected by the error. Hence many sign changes in divided differences are usual when the measurements are not smooth. However, if f is a k-times differentiable function whose k-th derivative has only a few sign changes, then the corresponding divided differences of exact function values also have only a few sign changes. Therefore it may be appropriate to modify the given measurements in order that there are few sign changes in the divided differences of the new values.

For example, if plotted values of the measurements show a single peak and away from the peak the underlying function seems to be constant or convex, then it would be suitable to introduce the condition that the second differences of the smoothed data change sign only twice.

In the general case k and q are given integers, and we require the sequence of k-th order divided differences of the smoothed data to change sign at most q times, where k and q are both less than n. We use the notation $\{y_i; i = 1,2,\ldots,n\}$ for the ordinates of the smoothed data whose abscissae are $\{x_i; i = 1,2,\ldots,n\}$. We regard the original measurements and the smoothed values as vectors, ϕ and y say, in $\mathbb{R}^n$. In order to define the "least" change to the data that gives the smoothness conditions, a norm is chosen in $\mathbb{R}^n$. The vector y is calculated to give the global minimum of $\|y-\phi\|$ subject to the conditions on the signs of its divided differences.

Three properties of this technique that may provide some useful advantages over other smoothing algorithms are as follows. There is no need to choose a set of approximating functions. The smoothing process is a projection because, if it is applied to y, then no changes are made to the components of y. By plotting the measurements it is often possible to identify appropriate values for k and q, or alternatively for a suitable k one can try increasing q until the differences $\{y_i-\phi_i;\ i = 1,2,\ldots,n\}$ seem to be due only to the errors of the measurements.

In the case when $q = 0$ and when the overall sign of the k-th divided differences is given, the constraints on the components $\{y_i;\ i = 1,2,\ldots,n\}$ are all linear. Therefore, if the ∞-norm or the 1-norm is chosen in $\mathbb{R}^n$, then the calculation of y is a linear programming problem, and, if the 2-norm is chosen, then a quadratic programming problem occurs. Thus several general algorithms are already available when $q = 0$ for the most frequently occurring norms.

Also, for certain values of k and q, there are some highly efficient special algorithms. In particular Section 2 considers methods of calculation in the monotonic case when $k = 1$ and $q = 0$. One method takes such a simple form when the ∞-norm is chosen in $\mathbb{R}^n$ that we are able to generalize it in Section 3 to the case when $k = 1$ and q is any non-negative integer. It is noted, however, that the generalization is not suitable for other norms because of the difficulties that are caused by isolated local solutions of the underlying minimization calculation.

Another interesting special case is when all second order divided differences are to be non-negative and the ∞-norm is chosen in $\mathbb{R}^n$. In this case $\{y_i;\ i = 1,2,\ldots,n\}$ can be obtained by adding a constant to the lower convex hull of the data $\{\phi_i;\ i = 1,2,\ldots,n\}$ (Ubhaya, 1977). This algorithm is generalized in Section 4 to allow q to be any non-negative integer.

In Section 5 some adverse results are given for the case when $k = 3$. We find that we no longer have the properties that are fundamental to the algorithms of Sections 3 and 4. In particular isolated local minima can occur for $q > 1$ even when the ∞-norm is chosen in $\mathbb{R}^n$. Moreover, if the smoothing gives the condition that the third differences of $\{y_i;\ i = 1,2,\ldots,n\}$ are non-negative, we find that it does not necessarily follow that there is a function g with a non-negative third derivative that satisfies the interpolation conditions $\{g(x_i) = y_i;\ i = 1,2,\ldots,n\}$.

In the final section there is a brief discussion of the given results and algorithms. In order to reduce the length of the paper, neither the propositions nor the efficacy of the algorithms are proved. Proofs will be given in the Ph.D. dissertation of one of the authors (MPC).

2. Monotonic function values

This section considers some algorithms for minimizing $\| y-\phi \|$ subject to the monotonicity conditions

$$y_1 \leqslant y_2 \leqslant \dots \leqslant y_n \, . \tag{2.1}$$

Except for some remarks on the ∞-norm that are made at the end of the section, we assume that minimizing $\| y-\phi \|$ is equivalent to minimizing a function of the form

$$F(y_1, y_2, \dots, y_n) = \sum_{i=1}^{n} d(y_i - \phi_i) \, , \tag{2.2}$$

where d is a real convex function whose least value is $d(0)$. In particular, if the L_p-norm is chosen, where $1 \leqslant p < \infty$, then d is the function

$$d(\theta) = |\theta|^p, \; -\infty < \theta < \infty \, . \tag{2.3}$$

It is convenient to break the required sequence of smoothed values $\{y_i; i = 1,2,\dots,n\}$ at the points where $y_{i-1} < y_i$, and to seek the intervals on which the components of y are constant. If s and t are any integers such that $1 \leqslant s < t \leqslant n+1$, we define Y_{st} to be the set of values of the real number η that minimize the expression

$$\sum_{i=s}^{t-1} d(\eta - \phi_i) \, . \tag{2.4}$$

It follows that, if $y_s = y_{s+1} = \dots = y_{t-1}$, if $s = 1$ or $y_{s-1} < y_s$, and if $t = n+1$ or $y_{t-1} < y_t$, then $y_s \in Y_{st}$. We note that the set Y_{st} is either a single point or a closed interval of the real line.

First we consider the case when the norm implies that there is only one real number, η_{st} say, in the set Y_{st} . It occurs when the function d of expression (2.2) is strictly convex, and it allows the following neat algorithm to be used to calculate the required smoothed values $\{y_i; i = 1,2,\dots,n\}$.

<u>Step 0</u> Set $s = 1$.

<u>Step 1</u> Calculate an integer t in $[s+1, n+1]$ such that η_{st} is the least of the numbers $\{\eta_{si}; i = s+1, s+2, \dots, n+1\}$. Set $y_i = \eta_{st}$ for $i = s, s+1, \dots, t-1$.

<u>Step 2</u> End the calculation if $t = n+1$. Otherwise increase s to t and go back to Step 1.

Most of the work of this algorithm is due to the large number of evaluations of η_{si} that are made in Step 1. Therefore the following sophisticated method, which is based on an idea in Kruskal (1964), is usually more efficient.

<u>Step 0</u> Set $s = 1$.

<u>Step 1</u> Set $t = s+1$ and $t^* = t$.

Step 2 If $t = n+1$ or if $\eta_{st} < \phi_t$ go to Step 4.

Step 3 Let ℓ be the greatest integer in $[t,n]$ such that $\eta_{st} \geq \phi_t \geq \phi_{t+1} \geq \ldots \geq \phi_\ell$. Replace t by $\ell+1$ and return to Step 2.

Step 4 If $t = t^*$ go to Step 8. Otherwise set $s^* = s$ and continue to Step 5.

Step 5 If $s = 1$ or if $y_{s-1} \leq \eta_{st}$ go to Step 7.

Step 6 Let ℓ be the least integer in $[1,s-1]$ such that $y_\ell = y_{\ell+1} = \ldots = y_{s-1} > \eta_{st}$. Replace s by ℓ and return to Step 5.

Step 7 If $s = s^*$ go to Step 8. Otherwise set $t^* = t$ and return to Step 2.

Step 8 Set $y_i = \eta_{st}$ for $i = s, s+1, \ldots, t-1$, which may reduce some values of y_i that were set earlier. End the calculation if $t = n+1$. Otherwise increase s to t and go back to Step 1.

We generalize the second of these algorithms to the case when the set Y_{st} may be a closed interval instead of a single point. It is an important case because it is particularly useful to work with the L_1-norm when there may be a few gross errors in the measurements $\{\phi_i;\ i = 1,2,\ldots,n\}$. In the generalization one gives η_{st} in Steps 2, 3, 5, 6 and 8 a specific value from Y_{st}, but the specific values have to be chosen carefully in order that the final y satisfies the monotonicity condition (2.1). We recommend defining η_{st} to be the greatest number in Y_{st} in Steps 2 and 3 and defining it to be the least number in Y_{st} in Steps 5, 6 and 8.

When the ∞-norm is chosen in $\mathbb{R}^n$, then the least value of $\|y-\phi\|$ subject to condition (2.1) is the number

$$\|y-\phi\|_\infty = \tfrac{1}{2} \max_{1 \leq i \leq j \leq n} (\phi_i - \phi_j). \tag{2.5}$$

It can be found in $O(n)$ computer operations, which provides several efficient algorithms for calculating optimal values of $\{y_i;\ i = 1,2,\ldots,n\}$.

3. Piecewise monotonic function values

Let $n = 10$ and let the measurements $\{\phi_i;\ i = 1,2,\ldots,n\}$ all have the values 0 except for $\phi_2 = \phi_9 = 1$. Consider the problem of calculating the vector y in $\mathbb{R}^{10}$ that minimizes $\|y-\phi\|_2$ subject to the conditions

$$\left.\begin{array}{ll} y_{i-1} \leq y_i, & i = 2,3,\ldots,j \\ y_i \geq y_{i+1}, & i = j,j+1,\ldots,n-1 \end{array}\right\}, \tag{3.1}$$

where the integer j can be regarded as one of the variables of the minimization calculation, and where the subscript 2 denotes the Euclidean vector norm. There are two solutions. One of them is $y_1 = 0, y_2 = 1, y_3 = y_4 = \dots = y_9 = 0.125$ and $y_{10} = 0$, and the other one may be obtained by exchanging y_2 and y_9. If the components of y have any other values that are allowed by the constraints (3.1), then $\|y-\phi\|_2$ is greater than its least value of $\sqrt{0.875}$. In particular for the best symmetric solution, namely $y_1 = y_{10} = 0$ and $y_2 = y_3 = \dots = y_9 = 0.25$, we have $\|y-\phi\|_2 = \sqrt{1.5}$.

If sufficiently small changes are made to the measurements in this example, then the optimal value of j is either 2 or 9. In order to identify the better value one would solve two separate quadratic programming problems. Therefore, when n is large and when several sign changes are allowed in the first divided differences of y, we expect the minimization of $\|y-\phi\|$ to require several optimization calculations in n variables. This conclusion is derived from the two separate solutions to the example, and similar behaviour can occur when the norm in $\mathbb{R}^n$ is any finite L_p-norm.

However, the purpose of this section is to show that, if the ∞-norm is chosen, then there are some fast algorithms for minimizing $\|y-\phi\|$ subject to piecewise monotonic conditions on y. In the case of the example the least value of $\|y-\phi\|_\infty$ is $\frac{1}{2}$ and it can be achieved by some vectors y that have the same symmetry as ϕ, and also some optimal vectors are similar to ones we had before, such as $y_1 = 0$, $y_2 = 1, y_3 = y_4 = \dots = y_9 = \frac{1}{2}, y_{10} = 0$. An important difference between the present case and the previous one is that now the set of optimal vectors y in $\mathbb{R}^{10}$ is connected. Therefore it does not follow that small perturbations to the measurements can cause local minima to occur in the optimization calculation that determines y.

Let Y be the set of vectors in $\mathbb{R}^n$ that minimize $\|y-\phi\|_\infty$ subject to at most q sign changes in the first divided differences of the components of y, and let y_a and y_b be any two elements of Y. Because Y is closed and bounded, there exist y_a^* and y_b^* in Y that minimize $\|y_a^*-y_b^*\|_\infty$ subject to the condition that, within Y, y_a^* and y_b^* are pathwise connected to y_a and y_b respectively. If $\|y_a^*-y_b^*\|_\infty$ is positive a contradiction can be obtained. Therefore the set of optimal vectors is connected in the general case, which gives some hope that efficient algorithms can be constructed for best piecewise monotonic approximations with respect to the ∞-norm.

First we consider the simplest case when y is constrained by the conditions (3.1). We let ℓ be any integer in $[1,n]$ that satisfies the equation

$$\phi_\ell = \max_{1\le i\le n} \phi_i . \tag{3.2}$$

By reducing any components of y that exceed y_ℓ to y_ℓ, it can be shown that it is optimal to set $j = \ell$. Hence the "up-down" approximation problem is reduced to two monotonic calculations.

Next we consider the "up-down-up" case. The key to the solution is to seek an improvement to the "up" case. It follows from the last paragraph of Section 2 that an improvement is obtained only if the "up-down-up" approximation decreases somewhere in $[x_s,x_t]$, where the integers s and t satisfy the equation

$$\max_{1\leqslant i\leqslant j\leqslant n} (\phi_i-\phi_j) = \phi_s-\phi_t . \tag{3.3}$$

Therefore the "up-down" part of the required approximation includes x_s and the "down-up" part includes x_t. It follows from the remarks of the last paragraph that we will have a best "up-down-up" approximation if we choose a best "up" approximation on $[x_1,x_s]$, a best "down" approximation on $[x_s,x_t]$, and a best "up" approximation on $[x_t,x_n]$, which again reduces the problem to monotonic calculations.

Finally we consider the case when there are $(q+1)$ sections on which the components of y are monotonic where $q \geqslant 3$. We suppose that the first section is "up". If $(q+1)$ is even, then the argument that uses equation (3.2) shows that we may let x_ℓ be a turning point of the final approximation, and, if $(q+1)$ is odd, then the argument that uses equation (3.3) implies that x_s and x_t may be assigned as final turning points. Hence the original range $[x_1,x_n]$ is divided into two or three sections, and the number of end-points that have not yet been chosen is even. For each section between assigned end-points we ask what the maximum error would be if the section were not sub-divided. Thus we find a section that would benefit most from sub-division. By applying to it the argument that depends on expression (3.3), we determine two more end-points of the final approximation. By continuing inductively, all the end-points are assigned, so again it is straightforward to reduce the piecewise monotonic problem to $(q+1)$ monotonic calculations.

4. Piecewise convex-concave smoothing

In the last section a brief argument supports the statement that the set of vectors y that minimize $\|y-\phi\|_\infty$, subject to piecewise monotonic conditions on the components of y, is connected. By a similar argument it may be shown that this property remains true when $k = 2$ and $q \geqslant 1$. Therefore it may be possible to find a fast method of calculating y to minimize $\|y-\phi\|_\infty$ subject to a given number of sign changes in the second differences of y. A suitable algorithm is given below.

In the usual case when there is some freedom in the optimal y, one has to decide on the particular approximation that one will calculate. The given algorithm causes y to satisfy the following conditions. The components y_1 and y_n satisfy the equations $|y_1-\phi_1| = |y_n-\phi_n| = \|y-\phi\|_\infty$, and their signs follow the given convexity-concavity conditions. For $i = 2,3,\ldots,n-1$, either the points $\{(x_{i-1}, y_{i-1}), (x_i,y_i), (x_{i+1},y_{i+1})\}$ are collinear, where we are using the usual co-ordinate notation for points in $\mathbb{R}^2$, or $|y_i-\phi_i| = \|y-\phi\|_\infty$. For example the solid line in

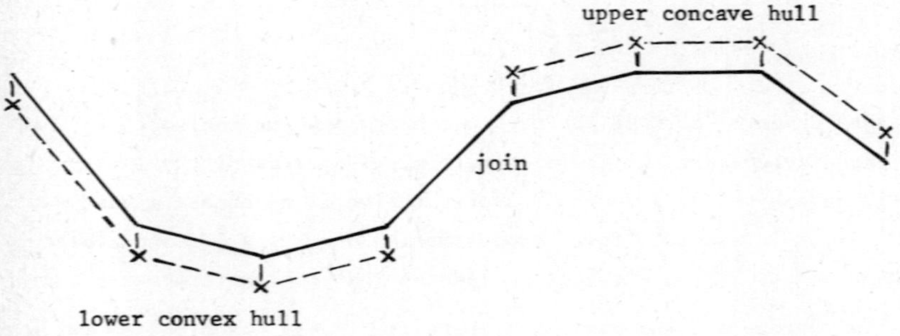

Figure 1. A best convex-concave approximation.

Figure 1 is the plot in $\mathbb{R}^2$ of the sequence of straight line segments joining $\{(x_i,y_i);\ i = 1,2,\ldots,n\}$, and the crosses give the positions of some of the data points $\{(x_i,\phi_i);\ i = 1,2,\ldots,n\}$. For each of the plotted data points the equation $|y_i-\phi_i| = \|y-\phi\|_\infty$ holds. There are some other data points that are not shown in the figure. In particular, in order to account for the maximum error $\|y-\phi\|_\infty$, there is at least one more point such that $|y_i-\phi_i| = \|y-\phi\|_\infty$, which is either above the plotted curve to the left of the join or below the plotted curve to the right of the join; we rule out the possibility that some data points inside the range of the "join" satisfy the equation $|y_i-\phi_i| = \|y-\phi\|_\infty$, by extending if necessary the "lower convex hull" or the "upper concave hull" to include such points.

First the algorithm is described in the case of Figure 1 when a convex-concave approximation is required. We find that it builds up the "lower convex hull" and the "upper concave hull" that are shown, working inwards from the ends of the range $[x_1,x_n]$. In Step 1 $L(s,t)$ and $U(s,t)$ are the lower and upper hulls respectively in $\mathbb{R}^2$ of the points whose co-ordinates have the values $\{(x_i,\phi_i);\ i = s, s+1, \ldots, t\}$. The number h is the greater of the minimax errors of two approximations, namely the best convex approximation to the data on $[x_1,x_s]$ and the best concave approximation to the data on $[x_t,x_n]$. Thus h is a lower bound on the final value of $\|y-\phi\|_\infty$, and, at the end of Step 1, h^* is an upper bound on $\|y-\phi\|_\infty$.

Step 0 Set $h = 0$, $s = 1$ and $t = n$.

Step 1 Calculate $L(s,t)$ and $U(s,t)$ and set h^* to half the greatest vertical distance between them. Increase h^* to h if $h > h^*$.

Step 2 If $h^* = 0$ go to Step 7. Otherwise continue to Step 3.

Step 3 Consider the straight line sections of $L(s,t)$, working from left to right. If the extension of any section passes below or through the point $(x_t, \phi_t - 2h^*)$, then add the section to the lower convex hull of the required approximation. Let (x_σ, ϕ_σ) be the last data point that is added to this lower convex hull.

Step 4 Consider the straight line sections of $U(s,t)$, working from right to left. If the extension of any section passes through or above the point $(x_s, \phi_s + 2h^*)$, then add the section to the upper concave hull of the required approximation. Let (x_τ, ϕ_τ) be the last data point that is added to this upper concave hull.

Step 5 If $\sigma = s$ and $\tau = t$ go to Step 7.

Step 6 Replace s and t by σ and τ respectively. Increase h if necessary so that it has the value that is defined in the paragraph that precedes the description of this algorithm. Go back to Step 1.

Step 7 The calculation of the convex and concave hulls is complete, and the optimal value of $\|y-\phi\|_\infty$ is h. The solid line of Figure 1, and hence $\{y_i;\ i = 1,2,\ldots,n\}$, is obtained by adding h to the lower convex hull, by subtracting h from the upper concave hull, and by letting the join be the straight line between the points $(x_s, \phi_s + h)$ and $(x_t, \phi_t - h)$.

Next we consider the convex-concave-convex case. We let the required approximation be the solid line that is shown in Figure 2. We extend the above algorithm

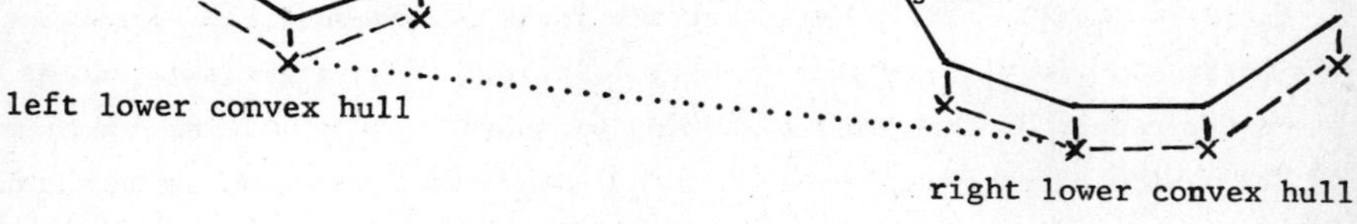

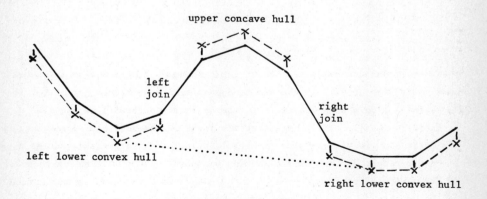

Figure 2. A best convex-concave-convex approximation.

to calculate the left and right lower convex hulls, the upper concave hull, and the value of $\|y-\phi\|_\infty$. Then it is straightforward to obtain the required approximation. The key to the calculation is that all sections of the lower convex hull of the data belong to the left lower or right lower convex hull, except for the dotted line in the figure. Therefore the best convex-concave-convex approximation is an improvement on the best convex approximation only if the largest error of the best convex approximation occurs in the range of the dotted line. Thus this range is found, we let (x_m,ϕ_m) be the data point that is furthest above the dotted line, and we note that it is in the upper concave hull of Figure 2.

If we split the range $[x_1,x_n]$ at x_m, then we have two approximation problems that are similar to the problem of Figure 1, except that a single value of $\|y-\phi\|_\infty$ is used to construct the required approximation from the three hulls of the two problems. Therefore an upper bound h^* on $\|y-\phi\|_\infty$ may allow the above algorithm to be used to generate some sections of the three hulls. Suitable modifications to the given algorithm should be clear, except perhaps for the following ones. At the beginning of the new version of Step 1 we have integers $s(1)$, $t(1)$, $s(2)$ and $t(2)$, such that the left lower convex hull has been found on $[x_1,x_{s(1)}]$, the upper concave hull has been found on $[x_{t(1)},x_{s(2)}]$, and the right lower convex hull has been found on $[x_{t(2)},x_n]$, where initially $t(1) = s(2) = m$. We let h be the greatest minimax error of three approximations, namely the best convex approximation to the data on $[x_1,x_{s(1)}]$, the best concave approximation to the data on $[x_{t(1)},x_{s(2)}]$, and the best convex approximation on $[x_{t(2)},x_n]$. We define h^* in Step 1 to be the maximum of the vertical distances between $L(s(1), t(1))$ and $U(s(1), t(1))$ and between $L(s(2), t(2))$ and $U(s(2), t(2))$, except that, as before, h^* is increased if necessary so that $h^* \geqslant h$. Moreover, using an obvious notation, Step 5 of the modified algorithm branches to Step 7 only if $\sigma(1) = s(1)$, $\tau(1) = t(1)$, $\sigma(2) = s(2)$ and $\tau(2) = t(2)$.

It is straightforward to generalize this algorithm to calculate a best approximation with q sign changes in the second divided differences from a best approximation with $(q-2)$ sign changes, provided that the signs of the initial second divided differences are the same. One makes use of the fact that all data points that are in convex and concave hulls of the first approximation are also in convex and concave hulls respectively of the second approximation. It does not matter if any hulls degenerate to a single point. At the beginning of Step 1 we require q pairs of integers $\{s(j), t(j)\}$ such that the joins of the required approximation are in the intervals $\{[x_{s(j)}, x_{t(j)}]\ ;\ j = 1,2,\ldots,q\}$. The algorithm reduces these intervals to the ranges of the joins by adding sections to lower convex and upper concave hulls. Reductions can occur in all the intervals, and when they are complete it is easy to construct y. Thus one can minimize $\|y-\phi\|_\infty$ subject to q sign changes in the second divided differences of y, where q is any non-negative integer.

5. Divided differences of order three

The algorithms that have been described for smoothing with respect to the ∞-norm depend on the following statement. If $k = 1$ or 2 and if y minimizes $\|y-\phi\|_\infty$ subject to a given sign of the k-th order divided differences of y, then we can find just $(k+1)$ of the measurements $\{\phi_i;\ i = 1,2,\ldots,n\}$ such that these measurements determine the optimal value of $\|y-\phi\|_\infty$. The first result of this section is that the statement may not be true when $k = 3$.

For example let $n = 7$, let $\{x_i;\ i = 1,2,\ldots,7\}$ and $\{\phi_i;\ i = 1,2,\ldots,7\}$ have the values $\{-5, -3, -1, 0, 1, 3, 5\}$ and $\{-12, -5, 0, 0, 0, 5, 12\}$ and let the third divided differences of y be non-negative. We seek the least value of $\|y-\phi\|_\infty$. By eliminating y_4 from the first and last divided difference conditions we obtain the inequality

$$-3y_1 + 10y_2 - 15y_3 + 15y_5 - 10y_6 + 3y_7 \geqslant 0 , \qquad (5.1)$$

which implies the bound $\|y-\phi\|_\infty \geqslant \frac{1}{2}$. This bound is satisfied as an equation if and only if $y_1 = -12\frac{1}{2}$, $y_2 = -4\frac{1}{2}$, $y_3 = -\frac{1}{2}$, $y_5 = \frac{1}{2}$, $y_6 = 4\frac{1}{2}$ and $y_7 = 12\frac{1}{2}$, and in this case we have non-negative third divided differences if and only if $y_4 = 0$. Therefore we have determined the optimal y. However, if we fit any four data by an approximation with a non-negative third divided difference, then we can achieve a minimax error that is less than $\frac{1}{2}$. Thus no set of four data determines the optimal value of $\|y-\phi\|_\infty$.

The main feature of the example is that the fourth data point is important, even though, because $y_4 = \phi_4$, the optimal vector y would not alter if ϕ_4 were changed by $\pm\frac{1}{2}$. In order to make this remark even clearer we suppose that the middle data point of the example is dropped. Then we can achieve $\|y-\phi\|_\infty = \frac{3}{8}$ without any negative third divided differences by letting the components of y have the values $-99/8$, $-37/8$, $-3/8$, $3/8$, $37/8$ and $99/8$.

This phenomenon would not occur if there existed a function g with a non-negative third derivative that satisfied the conditions $g(-5) = -99/8$, $g(-3) = -37/8$, $g(-1) = -3/8$, $g(1) = 3/8$, $g(3) = 37/8$ and $g(5) = 99/8$, because then all third divided differences of g would be non-negative, which would imply that we could let $y_i = g(x_i)$ in expression (5.1). However these values of y_i contradict condition (5.1). It follows that non-negative third divided differences of y do not imply that there exists a function g with a non-negative third derivative such that $\{g(x_i) = y_i;\ i = 1,2,\ldots,n\}$.

Another proof of this remark may be obtained from the expression for a divided difference that is given by the Peano kernel theorem (see Powell, 1981, for instance). The expression is the identity

$$g[x_j,\ x_{j+1},\ x_{j+2},\ x_{j+3}] = \int_{x_j}^{x_{j+3}} B_j(x) g'''(x)\,dx, \tag{5.2}$$

where B_j is a B-spline that is positive on (x_j, x_{j+3}) . Hence, if g''' is non-negative and if $g[x_1, x_2, x_3, x_4]$ and $g[x_3, x_4, x_5, x_6]$ are zero, it follows that g''' is zero almost everywhere on $[x_1,x_6]$, which implies that $g[x_2, x_3, x_4, x_5]$ is also zero. Therefore it is not possible for three consecutive third differences to have the values 0, 7 and 0, which occur in the example of the previous paragraph.

By giving detailed attention to expression (5.2), one can obtain conditions on non-negative third order divided differences that allow an interpolating function whose third derivative is non-negative. For example, if $n = 6$ and if the abscissae $\{x_i;\ i = 1,2,\ldots,6\}$ are equally spaced, then the condition is the inequality

$$\Delta^3 y_2 \leqslant \Delta^3 y_1 + \Delta^3 y_3 + 4\sqrt{\Delta^3 y_1 . \Delta^3 y_3}\ , \tag{5.3}$$

which is a nonlinear constraint on the components of y .

The final remark of this section is that, if we require the minimum of $\|y-\phi\|_\infty$ subject to at most two sign changes in the third divided differences of the components of y , then the set of optimal vectors y need not be connected. In order to demonstrate this point we let $n = 9$, and we let $\{x_i\}$ and $\{\phi_i\}$ have the values $\{-6, -5, -3, -1, 0, 1, 3, 5, 6\}$ and $\{-100, -12, -5, 0, 0, 0, 5, 12, 100\}$. The numbers ϕ_1 and ϕ_9 are such that we can restrict attention to the case when the first and last divided differences of y are positive. We assume connectedness and establish a contradiction.

The components $y_1 = -100$, $y_2 = -12$, $y_3 = -5$, $y_4 = -5/18$, $y_5 = -7/18$, $y_6 = 7/18$, $y_7 = 83/18$, $y_8 = 223/18$ and $y_9 = 100$ are optimal, but for the present argument it is only necessary to note that they are feasible and that $\|y-\phi\|_\infty = 7/18$. We let $\bar{y}$ be an optimal solution. Because of the symmetry of the data, the vector $\tilde{y}$ whose components are $\{\tilde{y}_i = -\bar{y}_{10-i};\ i = 1,2,\ldots,9\}$ is also optimal. Because the third divided differences of $\tilde{y}$ are the same as the third divided differences of $\bar{y}$, except that their order is reversed, we deduce from the connectedness of $\tilde{y}$ to $\bar{y}$ that there exists an optimal solution, $\hat{y}$ say, such that the least of its first three divided differences is equal to the least of its last three divided differences. If this least value is non-negative, then a simple extension to the example in the second paragraph of this section implies $\|\hat{y}-\phi\|_\infty = \frac{1}{2}$, which contradicts the value $\|y-\phi\|_\infty = 7/18$ that has been found already. Otherwise, because there are only two sign changes in the sequence of third differences, the values $\hat{y}[x_3,x_4,x_5,x_6]$ and $\hat{y}[x_4,x_5,x_6,x_7]$ are both negative. Therefore the sum of these terms gives the inequality

$$-\hat{y}_3 + 3\hat{y}_4 - 3\hat{y}_6 + \hat{y}_7 < 0 . \qquad (5.4)$$

It follows from the data that $\|\hat{y}-\phi\|_\infty > 5/4$, which is also a contradiction. Therefore local minima can occur in the minimization of $\|y-\phi\|_\infty$ when $k = 3$ and $q \geqslant 2$.

6. Discussion

The given algorithms solve problems that have the nice property that any local solution of the constrained minimization calculation is also a global minimum. However, these algorithms can be applied only if no sign changes are allowed in the k-th divided differences of y , or if the ∞-norm is chosen and the order of the divided differences is at most two. These restrictions rule out most of the useful applications of the smoothing method.

Therefore some research on the difficulties that are caused by local minima may be valuable. The fact that the total number of local minima is finite may be an advantage, and some special cases have properties that are very helpful indeed. For instance, if the least value of $\|y-\phi\|$ is required subject to condition (3.1), where j is still a variable and where the norm is general, then ϕ_j is a local maximum value in the sequence $\{\phi_i;\ i = 1,2,\ldots,n\}$ and one can let $y_j = \phi_j$.

The remarks on third divided differences in Section 5 suggest that, instead of the general smoothing technique that is proposed, it may be better to minimize $\|y-\phi\|$ subject to the condition that the smoothed values can be interpolated by a function whose k-th derivative changes sign at most q times. A drawback of this approach is that it involves nonlinear constraints on the components of y that are at least as elaborate as inequality (5.3).

Little use has been made so far of electronic calculating machines to test the smoothing technique.

References

Hildebrand, F.B. (1956), Introduction to numerical analysis, McGraw-Hill (New York).

Kruskal, J.B. (1964), "Nonmetric multidimensional scaling: a numerical method", Psychometrika, Vol. 29, pp. 115-129.

Powell, M.J.D. (1981), Approximation theory and methods, Cambridge University Press (Cambridge).

Ubhaya, V.A. (1977), "An O(n) algorithm for discrete n-point convex approximation with applications to continuous case", Technical Memorandum No. 434 (Operations Research Dept., Case Western Reserve University).

ON THE CONTROL OF THE GLOBAL ERROR IN STIFF INITIAL VALUE PROBLEMS

Germund Dahlquist

Summary. An approach of B. Lindberg [3] to the estimation and control of a norm of the global error is modified and applied to systems of stiff ODE's in partitioned form.

1. Introduction

Most packages for the solution of initial value problems control the step size and order with the aid of a criterion on the local error. The relation of a parameter, sometimes called TOL, to the global error is often very obscure. Some packages provide global error estimates for non-stiff problems.

Lindberg [3] has given a characterization of the optimal step size sequences, which is the basis of the present paper. Roughly speaking, he shows that in typical stiff problems one should keep the global error at the maximum level during large intervals. Before it has reached that level a quantity that may be described as "the discounted value" of the local truncation error per step should be kept constant. To obtain the value of this constant requires, however, some forecasting. When we planned this study, we judged, perhaps erroneously, that this forecast would be too unreliable. With our control strategy the global error tends more slowly to its maximum level, see Section 3. We need an estimate of the local rate of decay (growth) of the previously committed errors, which is a central concept also in Lindberg's theory. Our scheme appears to work well on dissipative problems. The suggestions are more tentative for non-dissipative problems and for problems where the growth rate is increasing.

This is a report on work in progress. It is more a contribution to a discussion then a definite recommendation for practical use. Some numerical experiments have been done but much more work, experimental as well as theoretical, remains to be done.

2. Proposition for a step size control

Let z(t) be the solution of an initial value problem for s simultaneous ODE's,

$$S\,dz/dt = F(t,z), \tag{2.1}$$

where S is a constant diagonal matrix, $S > 0$. Set

$$J(t,u,v) = \int_0^1 F'(t,u+\theta v)\,d\theta, \tag{2.2'}$$

and note that

$$F(t,z+v) - F(t,z) = J(t,z,v)\cdot v. \tag{2.2''}$$

Let r(t) be the local truncation error per unit of time of the numerical method under consideration, and consider the "pseudo-linear" system,

$$S\,dw/dt - J(t,z(t),w)w = -Sr(t). \tag{2.3}$$

The global errors of our numerical method satisfy a difference equation, approximately the one which is obtained when the same method is applied to (2.3). One of our basic assumptions is the following:

ASSUMPTION I: The sequence of global errors is well described by the solution w(t) of (2.3). ■

This is the only point where the stability properties of the numerical method enter in our study. In fact, the choices of step size and step size ratio suggested by the theory of this paper have to be examined by means of simple sufficient stability criteria for the method. It is an object of the stability theory of the numerical method to provide such criteria, but it is beyond the scope of this paper.

Our step size control will be based on (2.3). Let $\langle u,v\rangle$ be an inner-product in $\mathbb{R}^s$ and let $\|u\|^2 = \langle u,u\rangle$. By (2.3),

$$\langle w, S\,dw/dt\rangle = \langle w, J(t,\hat{z}(t),w)w\rangle - \langle w, Sr(t)\rangle . \tag{2.4'}$$

ASSUMPTION II: We can compute a function m(t), such that

$$\langle w, J(t,\hat{z}(t),w)w\rangle \lesssim m(t)\,\langle w, Sw\rangle \quad \text{for} \quad w = w(t).$$

■

We shall return to this in Section 4.

Put

$$\langle x, Sx\rangle = \|x\|_S^2, \tag{2.4''}$$

and note that

$$\langle w, S\,dw/dt\rangle = \|w\|_S \cdot d\|w\|_S/dt.$$

Apply the Schwarz inequality to the last term of (2.4'), and divide by $\|w\|_S$. Then

$$d\|w\|_S/dt \lesssim m(t)\|w\|_S + \|r(t)\|_S. \tag{2.5}$$

Our estimate of the global error is obtained by an application of the implicit Euler method to this inequality.

ASSUMPTION III. The control strategy depends on a function $\hat{r}(t)$ which is assumed to be chosen so that there exists a step size such that $\|r(t)\|_S \le \hat{r}(t)$. ■

Then $\|w(t)\|_S \le \bar{w}(t)$, where $\bar{w}(t)$ is the solution of our basic *error norm equation*:

$$d\bar{w}(t)/dt = m(t)\bar{w}(t) + \hat{r}(t). \tag{2.6}$$

ASSUMPTION IV. The requirement on the global error is expressed in the form

$$\bar{w}(t) \le \hat{w}(t), \quad \forall t \in [t_0,T]. \tag{2.7}$$

(There are several reasons to allow $\hat{w}$ to depend on t, see Section 5.)

■

Set

$$v(t) = \bar{w}(t)/\hat{w}(t). \qquad (2.8')$$

Then by (2.6), (2.7),

$$dv/dt = \bar{m}(t)v + \bar{r}(t), \quad v(t) \le 1, \qquad (2.8'')$$

where

$$\begin{aligned} \bar{m}(t) &= m(t) - d\ln\hat{w}/dt \\ \bar{r}(t) &= \hat{r}(t)/\hat{w}(t) \end{aligned} \qquad (2.8''')$$

Eqn. (2.8") is called the *scaled error norm equation*.

In order to define a strategy we first consider the following simplified problem: Given $t' \in [0,T[$, $v' \in [0,1[$, and assume that $\bar{m}$ is constant, how shall we choose a constant $\bar{r}$ so that there is a solution $v^*(t)$, to the problem

$$dv^*/dt = \bar{m}v^* + \bar{r}, \quad t \in [t',T],$$
$$v^*(t') = v', \; v^*(T) = 1.$$

It is easily verified that, for $\bar{m} \neq 0$,

$$v^*(t) = v'\exp\bar{m}(t-t') - (r/\bar{m})1-\exp\bar{m}(t-t'))$$

and hence $v^*(T) = 1$ iff

$$\bar{r} = -\bar{m}\cdot\frac{1-v'\exp(\bar{m}(T-t'))}{1-\exp\bar{m}(T-t')}.$$

Our control is, in the case of a variable $\bar{m}$ chosen in accordance with this, with a safety factor $1-\xi$, i.e.

$$\bar{r}(t) = \begin{cases} -(1-\xi)\bar{m}(t)\dfrac{1-v(t)\exp(\bar{m}(t)(T-t))}{1-\exp(\bar{m}(t)(T-t))}, & \bar{m}(t) \neq 0, \\[2ex] \dfrac{1-v(t)}{T-t}, & \bar{m}(t) = 0. \end{cases} \qquad (2.9)$$

In the following discussion we assume that $\xi=0$. By (2.8")

$$dv/dt = \bar{m}(t)v + \bar{r}(t) = m_1(t)(v(t) - 1) \qquad (2.10')$$

where

$$m_1(t) = \bar{m}(t)/[1-\exp(\bar{m}(t)(T-t))]. \qquad (2.10'')$$

Note that $-\infty < m_1(t) < 0$ for $0 < t < T$. By integration of (2.10'),

$$\ln|1-v(t)| = \ln|1-v(t')| + \int_{t'}^{t} m_1(\tau)d\tau. \qquad (2.10''')$$

It follows that if $v(t') < 1$, then $v(t') < v(t) < 1$ for $t \in [t',T[$ and, by (2.10'), $dv/dt > 0$, i.e. $v(t)$ increases.

Note that if $\exp(m(t)(T-t)) \ll 1$, then by (2.9),

$$\bar{r}(t) \approx -(1-\xi)\bar{m}(t),$$

i.e.

$$\hat{r}(t) \approx -(1-\xi)\bar{m}(t)\hat{w}(t) \text{ if } \exp\bar{m}(t)(T-t) \ll 1. \qquad (2.11)$$

By (2.6) this means that $\bar{w}(t)$ will rapidly approach $\hat{w}$ in this case (if $\xi=0$), see also Fig.1, which is in agreement with the Lindberg theory [3].

The following result is obtained after a short calculation based on (2.10‴) and (2.10').

LEMMA. If $\bar{m}(t)$ is constant, then $\bar{r}(t)$ is also a constant, (which depends on $\bar{m}$), i.e. we can put t=0 (or any constant value instead of t) into the right hand sides of (2.9).

■

Fig.1 shows v(t) for $t \in [t',T]$ for different values of the constant $\alpha = \bar{m}\cdot(T-t')$. If $\bar{m}(t)$ is variable, say that $\alpha_0 \le \bar{m}(T-t') \le \alpha_1$, v(t) will be contained between the curves for $\alpha = \alpha_0$ and $\alpha = \alpha_1$. Fig.2 shows $(T-t')\bar{r}$ as a function of α for v' = 0,0.5,1,2.

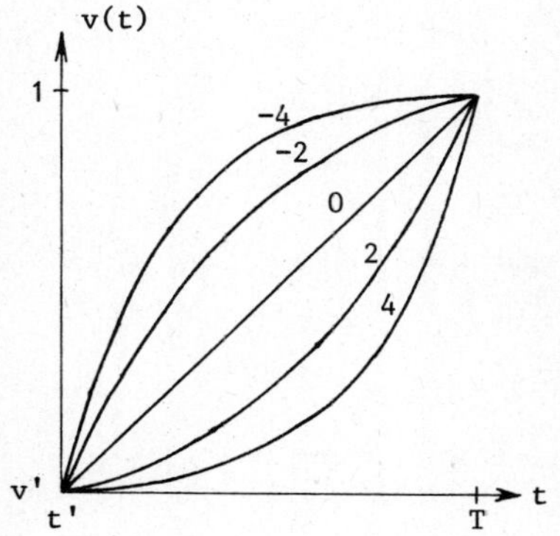

Fig.1

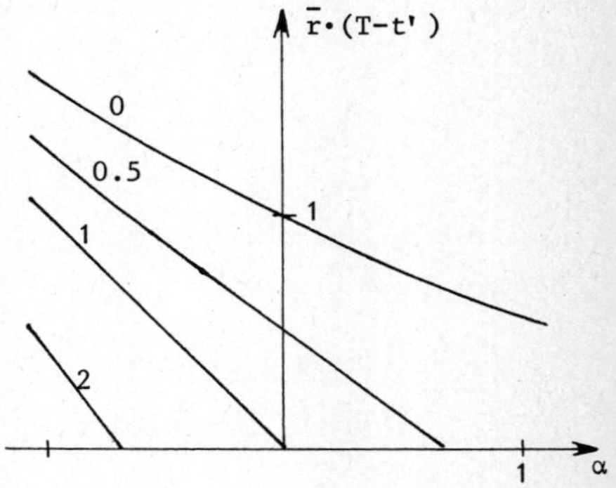

Fig.2

Eqn.(2.9) makes no sense unless $\bar{r}(t) > 0$. If $\bar{m}(t) \le 0$ everywhere there is no trouble, because then $v(t) < 1$ and hence $\bar{r}(t) > 0$. If $\bar{m}(t) = \bar{m} =$ const. for $t \ge t'$, we know, by the lemma, that $\bar{r}(t) \equiv \bar{r}(t')$ which is positive, if $v(t')\exp \bar{m}(T-t') < 1$. This is true also if $\bar{m} > 0$.

One can also show that $\bar{r}(t)$ will never become negative, when $\bar{m}(t)$ is non-increasing.

Hence there is a risk that $\bar{r}(t)$ becomes negative only when $\bar{m}(t)$ is positive and increasing. Our way of handling such ill-conditioned cases is tentative only. We allow the function $\hat{w}(t)$ (see Assumption IV) to increase, when $v(t')\exp m(t')(T-t') = 0.8$ (say), setting

$$d \ln \hat{w}(t)/dt = m(t) - m(t'),$$

for t>t' as long as $m(t) \ge m(t')$. By (2.8‴) this means that $\bar{m}(t) = m(t') =$ const. when $t>t'$, $m(t) \ge m(t')$. Hence $\bar{r}(t)$ remains positive, but the tolerance for the global error originally set will be exceeded.

3. The efficiency of the proposed strategy

Consider the problem (2.8") and assume that $\hat{w} = \text{const.}$ and that the method is such that we can set

$$\bar{r}(t) = h^p(t)\cdot|y^{p+1}(t)|c/\hat{w}. \tag{3.1}$$

We assume that the step size is changed at every step and make a continuous imbedding of the step size process, i.e. we interpolate a continuous piecewise differentiable function h(t) through the actual step sizes h_n and set e.g.

$$(h_{n+1} - h_n)/h_n = (h_{n+1} - h_n)/(t_{n+1} - t_n) \approx dh/dt \tag{3.2}$$

see also Lindberg [3].

Lindberg, see p.861 and 875, introduces,

$$\beta(t) = \rho(t)v(t), \; \rho(t) = \exp - \int_0^t \bar{m}(s)ds. \tag{3.3}$$

Then (2.8") reads

$$\dot{\beta}(t) = \bar{r}(t)\rho(t), \; \beta(0) = 0, \; \beta(t) \le \rho(t). \tag{3.4}$$

The functions $\beta(t)$ and $\bar{r}(t)\rho(t)$ may be described as "the discounted value" (at t=0) of, respectively, v(t) and the scaled local error *per unit of time*.

A corollary of Lindberg's theorem:
The function h(t), which minimizes the total number of steps,

$$N = \int_0^T dt/h(t), \tag{3.5}$$

subject to the constraints given by (3.4), (3.1), is a continuous function such that $v(t) = 1$ in some sub-intervals. The set $\{t:v(t)<1\}$ may consist of one or several intervals. The discounted value of the local truncation error *per step*, i.e. $h(t)\bar{r}(t)\rho(t)$, is constant on each of these sub-intervals.

PROOF: Lindberg sets,

$$\eta(t) = h^{-p}\bar{r}(t)\rho(t) \tag{3.6}$$

and shows, see statement (iii) of Theorem 1, that if $\beta(t) \neq \rho(t)$ on a sub-interval I then,

$$\dot{\beta}(t) = c(I)\cdot[\eta(t)]^{1/(p+1)}, \; t \in I.$$

Then, by (3.4) and (3.6),

$$h^p\eta(t) = c(I)\cdot[\eta(t)]^{1/(p+1)},$$

hence

$$h^{p+1}\eta(t) = [c(I)]^{(p+1)/p} = \text{const. for } t \in I.$$

■

We shall now compare this optimal step size sequence with the strategy proposed in Section 2, idealized in the sense that we think of a continuous model with perfect estimates of the local error. We first consider the model problem,

$$y' = -y, \;\; y(0) = 1, \;\; T = \infty, \tag{3.7}$$

with constant tolerance $\hat{w}$, constant order of consistency p, and $\xi=0$. For more general equations

$$y' = my, \quad m = \text{const.} < 0, \tag{3.8}$$

we only need to replace t, h etc. by, respectively, $|mt|$, $|mh|$ etc. The estimate of the total number of steps (3.5) is independent of m. For (3.7),

$$\rho(t) = e^{t}, \ \eta(t) = c/\hat{w}, \ \bar{r}(t) = h^{p}e^{-t}c/\hat{w}. \tag{3.9}$$

Lindberg's theorem then gives $h = \text{const.}\ h = h_0$ (say) as long as $v(t) < 1$. Then, for $t \geq t^*$ (say), $v(t) = 1$, i.e.

$$\dot{\beta}(t) = h_0^{p}c/\hat{w}, \quad t \leq t^*,$$

$$\beta(t) = \exp t, \quad t \geq t^*.$$

The continuity of $\dot{\beta}$ and β at $t = t^*$ then yields,

$$t^* = 1,$$

$$h_0^{p}c/\hat{w} = \exp t^* = e.$$

For $t \geq 1$ we have $v(t) = 1$ and hence, by (2.8"), $\bar{r}(t) = 1$. For $t \leq 1$, $\bar{r}(t) = \exp(1-t)$, by (3.9). By (3.9) and (2.8") we then obtain, for this *optimal* case,

$$\text{For } t \leq 1: \quad h_{opt}^{p}(t) = e\hat{w}/c, \quad v_{opt}(t) = te^{1-t}, \tag{3.10'}$$

$$\text{For } t \leq 1: \quad h_{opt}^{p}(t) = e^{t}\hat{w}/c, \quad v_{opt}(t) = 1. \tag{3.10''}$$

The number of steps in the optimal sequence is

$$N_{opt} = \int_0^{\infty} dt/h_{opt}(t) = (c/\hat{w})^{1/p}(1+p)\exp(-1/p).$$

For *the proposed* strategy we obtain, by the Lemma of Section 2, in this model problem

$$\forall t, \ \bar{r}_{prop}(t) = 1.$$

Then, by (3.9) and (2.8")

$$\forall t, \ h_{prop}^{p}(t) = e^{t}\hat{w}/c, \quad v_{prop}(t) = 1 - e^{-t}, \tag{3.11}$$

$$N_{prop} = \int_0^{\infty} dt/h_{prop}(t) = (c/\hat{w})^{1/p}\cdot p.$$

Hence

$$N_{prop}/N_{opt} = (1 + 1/p)^{-1} \exp(1/p). \tag{3.12}$$

We obtain, for the BDF methods, $c = 1/(p+1)$,

	2	3	4	5	6
N_{prop}/N_{opt}	1.099	1.047	1.027	1.018	1.012
N_{prop} for $\hat{w}=10^{-4}$	115	40.7	26.8	22.0	20.1

Another interesting fact is illustrated by this test problem. For the proposed strategy we have, by (3.11)

$$dh/dt = h/p,$$

and, when $t \geq 1$, this is true also for the optimal step size sequence. By (3.2), we obtain,

$$(h_{n+1}/h_n)_{prop} \approx 1 + h_n/p$$

or, in terms of Eqn. (3.8),

$$(h_{n+1}/h_n)_{prop} \approx 1 + |m|h_n/p. \tag{3.13}$$

This shows that the step size ratios are not very large, as long as the solution of (3.8) is still alive. When $y(t) = \hat{w}$ we obtain, by (3.11), $|mh_n| = c^{-1/p}$. The step size ratios are, however, sufficiently large to be able to cause instability with BDF-methods of order 6 or 5, towards the end of a transient of a stiff problem, see [4]. If there is a big gap between two successive time constants, there may be an interval where a rapid increase of step size is asked for. When this happens it may be advisable to switch to one of the two-step methods derived in [2], which are A-contractive for any step size ratio.

Our method of control cannot be expected to work equally well in a problem without dissipation, but we shall see, that it is not too bad except in fairly extreme cases. Consider (2.6), (2.7) with $m(t) = 0$, e.g. a quadrature problem.

Then the proposed strategy leads to constant local error *per unit of time*. Hence,

$$ch^p|y^{(p+1)}| = \hat{w}/T,$$

$$N_{prop} = \int_0^T dt/h(t) = (cT/\hat{w})^{1/p} \int_0^T |y^{(p+1)}|^{1/p} dt = \left(cT^{p+1}/\hat{w}\right)^{1/p} \|y^{(p+1)}\|_{1/p}^{1/p}$$

where we have set

$$\|f\|_q = \left(\frac{1}{T}\int_0^T |f|^q dt\right)^{1/q} \tag{3.14}$$

It has, however, been known for a long time, see e.g. [5, p.104] that the optimal step size sequence has constant local error *per step*, i.e.

$$c\,h^{p+1}|y^{(p+1)}| = \hat{w}/N_{opt}$$

$$N_{opt} = \int_0^T dt/h(t) = \left(T^{p+1}\, c\, N_{opt}/\hat{w}\right)^{1/(p+1)} \|y^{(p+1)}\|_{1/(p+1)}^{1/(p+1)}.$$

After a short calculation we obtain,

$$N_{opt} = (cT^{p+1}/\hat{w})^{1/p} \|y^{(p+1)}\|_{1/(p+1)}^{1/p}.$$

Hence if $m(t) \equiv 0$, then

$$N_{prop}/N_{opt} = \left(\|f\|_{1/p}/\|f\|_{1/(p+1)}\right)^{1/p}, \quad f = y^{(p+1)}. \tag{3.15}$$

We take two examples of integrals over an infinite interval, translated where the remainder is equal to $\hat{w}$.

EXAMPLE: Take $y^{(p+1)}(t) = e^{-t}$, $T = -\ln \hat{w}$. For $\hat{w} = 10^{-4}$, $p = 4$ we obtain $N_{prop}/N_{opt} = 1.04$, a rather good result.

The ratio is larger if $|y^{(p+1)}(t)|$ is very much smaller than its average in a very large part of the interval.

EXAMPLE: Take $y(t) = (1+t)^{-2}$, $T+1 = 1/\hat{w}$. Then $y^{(p+1)}(t) = c(1+t)^{-3-p}$. For $\hat{w} = 10^{-4}$, $p = 4$, we obtain $N_{prop}/N_{opt} = 4.4$. This case is, however, rather extreme. The use of the continuous model, and even the use of asymptotic error estimates, is more than doubtful here. In a less extreme case e.g. for $T = 9$, $\hat{w} = 10^{-4}$, $p = 4$ we obtain $N_{prop}/N_{opt} = 1.04$.

4. The estimation of m(t)

We shall consider partitioned systems. Set, in (2.1) and (2.3),

$$z = \begin{bmatrix} x \\ y \end{bmatrix}, \quad F = \begin{bmatrix} f \\ g \end{bmatrix}, \quad S = \begin{bmatrix} I & 0 \\ 0 & E \end{bmatrix}, \quad w = \begin{bmatrix} u \\ v \end{bmatrix}. \tag{4.1}$$

We then write (2.1) and (2.3) in the form,

$$\begin{aligned} \dot{x} &= f(t,x,y), \\ E\dot{y} &= g(t,x,y), \quad E = \mathrm{diag}(\varepsilon_1, \varepsilon_2, \ldots, \varepsilon_r), \quad \varepsilon_i \geq 0, \end{aligned} \tag{4.2}$$

$$\begin{aligned} \dot{u} &= J_{11}u + J_{12}v - r_x, \qquad u(0) = 0, \\ E\dot{v} &= J_{21}u + J_{22}v - Er_y, \qquad v(0) = 0, \end{aligned} \tag{4.3}$$

r_x, r_y are the local truncation errors per unit of time. We assume that the scaling is done so that the matrices J_{ij} and J_{22}^{-1} are bounded, uniformly with respect to the ε_i.

Let us temporarily set $E = \varepsilon I$, $0 < \varepsilon << 1$. *In the transient*, set

$$t = \varepsilon\tau, \quad r_x = r_x^0/\varepsilon, \quad r_y = r_y^0/\varepsilon.$$

Then

$$\begin{aligned} dx/d\tau &= \varepsilon f(\varepsilon\tau,x,y), \qquad du/d\tau = \varepsilon J_{11}u + \varepsilon J_{12}v - r_x^0 \\ dy/d\tau &= g(\varepsilon\tau,x,y), \qquad dv/d\tau = J_{21}u + J_{22}v - r_y^0. \end{aligned}$$

By induction,

$$\begin{aligned} d^v x/d\tau^v &= O(\varepsilon) \Rightarrow r_x^0 = O(\varepsilon\cdot\Delta\tau^p) \Rightarrow u(\tau) = O(\varepsilon\Delta\tau^p) \\ d^v y/d\tau^v &= O(1) \Rightarrow r_y^0 = O(1\cdot\Delta\tau^p) \Rightarrow v(\tau) = O(\Delta\tau^p). \end{aligned}$$

Hence, *in the transient*,

$$u(\tau) = O(\varepsilon\|v(\tau)\|). \tag{4.4}$$

In stiff intervals, we obtain by (4.3),

$$v = -J_{22}^{-1}J_{21}u + O(\varepsilon\Delta t^p). \tag{4.5}$$

Note that this means that the global error of v is essentially a follower of that of u, and it has very little to do with r_y. If we introduce (4.5) into (4.3), we obtain

$$\dot{u} = J_s u - r_x + O(\varepsilon\Delta t^p), \tag{4.6'}$$

where J_s is the Schur complement, which is independent of the choice of E,

$$J_s = J_{11} - J_{12}J_{22}^{-1}J_{21}. \tag{4.6''}$$

Note that u(t) and v(t) are both $O(\Delta t^p)$ in stiff intervals, and that u(t) is the important error to be controlled, in stiff intervals. Let the inner-product defined in Section 2 be of the form,

$$\langle w',w''\rangle = \langle u',u''\rangle + \langle v',v''\rangle .$$

Then the S-norm, see (2.4"), becomes

$$\|w\|_S^2 = \langle w,Sw\rangle = \langle u,u\rangle + \langle v,Ev\rangle .$$

Note that, by (4.4) and (4.5)

$$\langle v,Ev\rangle / \langle u,u\rangle = \begin{cases} O(1/\varepsilon) & \text{in the transient} \\ O(\varepsilon) & \text{in stiff intervals.} \end{cases} \tag{4.7}$$

Hence, *in the transient v gives a strongly dominant contribution to* $\|w\|_S$, *while in the stiff intervals the dominant contribution to* $\|w\|_S$ *is given by u.*

If we could compute

$$m(t) = \frac{\langle w,\ J(t,\hat{z}(t),w)w\rangle}{\langle w,Sw\rangle} \tag{4.8}$$

with w equal to the true global error vector w(t), then we would have equality in (2.5), apart from the use of the Cauchy-Schwarz inequality in the last term of (2.4') Therefore the choice of inner-product norm would not be very critical. (When one asks for rigorous error bounds, one sometimes needs the supremum of the right hand side of (4.8). In that context the choice of norm is more critical.)

Unfortunately, we cannot compute m(t) in this way, for two reasons:

1) we don't work with the vector w(t), we have only an estimate of its S-norm;
2) the Jacobian is not available for the whole system in our program for handling partitioned systems.

We therefore need a good substitute for w, $\tilde{w} = (\tilde{u},\tilde{v})$ and compute, for some η, which should be small, though not too small,

$$m(t) = \frac{\langle \tilde{u},f(z+\eta\tilde{w}) - f(z)\rangle + \langle \tilde{v},g(z+\eta\tilde{w}) - g(z)\rangle}{\eta(\langle \tilde{u},\tilde{u}\rangle + \langle \tilde{v},E\tilde{v}\rangle)} \tag{4.9}$$

The demand for accuracy in m(t) is not very high, when $\bar{m}(t) \le 0$. Since $\bar{r}(t)$ is proportional to h^p, we have, by (2.11) or (2.9), the following relation between perturbations in $\bar{m}$ and h:

$$\frac{\delta h}{h} \approx \frac{1}{p} \cdot \frac{\delta\bar{m}}{\bar{m}} \qquad (\text{if } \bar{m}(t) < 0).$$

It is, for example, not necessary to compute m(t) at every step. Roughly speaking, the spectral composition of $\tilde{w}$ should resemble that of w. An over-estimate of m(t) is safer than an underestimate, since it results in a smaller step size and an over-estimate of the global error.

In the transient we use second order backward differences,

$$\tilde{u} = \underline{D}^2x, \quad \tilde{v} = \underline{D}^2y. \tag{4.10}$$

We tried $D_-^p z$, but the oscillations in the step size sequence produced sometimes became disturbing. The results of the use of (4.10) seem rather satisfactory, see also [3], Section 5. We are likely to obtain overestimates, when we use differences of too low order.

In stiff intervals, we believe that the relation (4.5) should be approximately satisfied. Since a similar relation does not always hold between $\ddot{x}$ and $\ddot{y}$, we choose $\tilde{u}, \tilde{v}$ so that

$$\tilde{u} = D_-^2 x, \quad J_{22}\tilde{v} \approx -[g(z+\eta\tilde{u}) - g(z)]/\eta. \tag{4.11}$$

We can use the available approximate factorization of $\alpha E/h - J_{22}$ here to obtain $\tilde{v}$. Note that (4.11) means that

$$m(t) \approx \langle \tilde{u}, J_s \tilde{u} \rangle / \langle \tilde{u}, \tilde{u} \rangle,$$

as it should, according to (4.6'). According to (4.7), the ratio $\langle v, Ev \rangle / \langle u, u \rangle$ can be used for the switching between transient and stiff interval.

Some modification is done to the step size suggested by this strategy, partly to avoid oscillations in the step size sequence, partly to guarantee numerical stability. Since we use the Jacobian only in stiff intervals and then for the lower part of the system only, it may not be necessary to keep h piecewise constant to avoid the refactorization of the Jacobian, at least not when $\|E\| \ll h$.

5. On the use of a time-dependent $\hat{w}$

There are at least three reasons for the consideration of a time dependent tolerance function $\hat{w}(t)$:

a) The accuracy specification given by the user may be expressed in another norm than the S-norm internally used, e.g. in some weighted ℓ_2-norm. Note that, by (4.7), $\|w\|_S^2 / \|w\|^2$ is $O(\varepsilon)$ in the transient and $O(1)$ in the stiff intervals.

b) The user may be satisfied with a less stringent tolerance in the transient phase than during the later part of the computations. Note, however, that, in non-linear problems, the transient must be computed with sufficient accuracy to make sure that the solution does converge to the correct pseudo-steady state. In problems where the solution tends to a steady state, it is often both interesting and possible to determine this state with a higher accuracy than is needed in the earlier stages.

c) The initial value problem (for the differential system itself) may be more ill-conditioned than anticipated. In such a case the program must be able to switch from an absolute tolerance to a tolerance related to the growth of the solution of the variational equation.

Such cases are covered by our theory, if one lets $\hat{w}$ depend on t. One has to change m(t) and r(t) according to (2.8'''), as indicated at the end of Section 2. We omit the details.

6. A numerical example

The following system was solved by the BDF-method, variable order ≤ 4, with the proposed strategy, with $\hat{w} = 10^{-4}$, $\xi = 0.5$. The ℓ_2-norm was used.

$$\dot{x}_1 = 0.6 + x_2 - 0.8x_1 - 0.536x_1^2, \quad x_1(0) = 0$$
$$\dot{x}_2 = 3.48x_1 - x_2 \quad , \quad x_2(0) = 0.$$

This system is obtained from the equations for the oxidation of propane, eqn. (2.3.3) of [1], $\varepsilon = 1$, keeping the non-stiff variables constant. Fig.3 displays the variation of x_1, x_2, h, m and v, where v = estimated global error. The order was equal to 4 almost all the time. The total number of steps until t = 12.88 was N = 100.

Some sample values of m(t) and the eigenvalues of the Jacobian:

x_1	m(t)	λ_1	λ_2
0	-1.9	+0.97	-2.77
2.5	-3.0	0	-4.48
5.1	-0.4	-0.41	-6.86

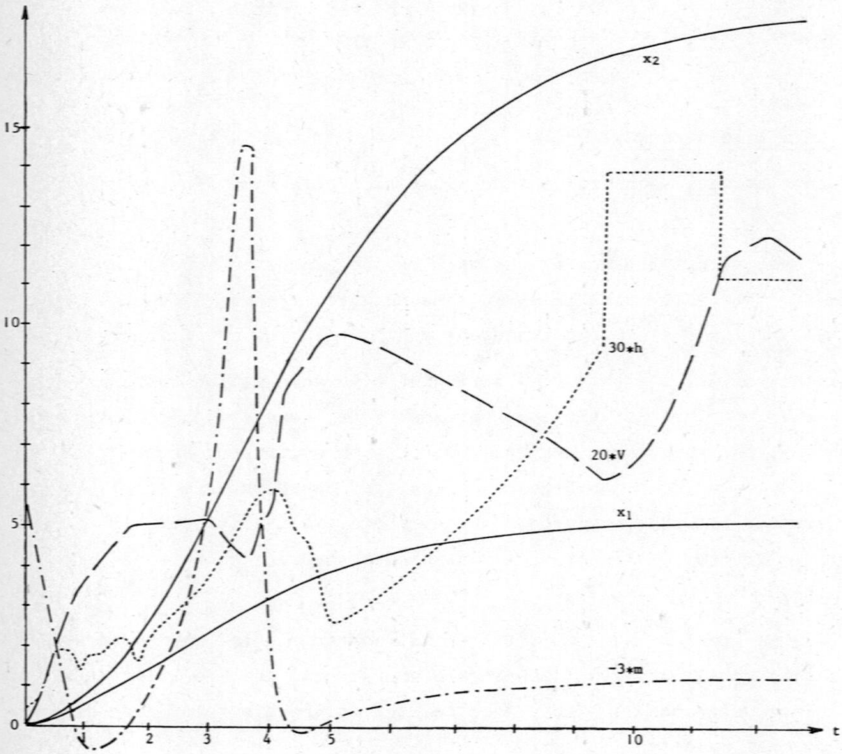

Fig.3

Acknowledgement

The author wants to acknowledge the assistance of Mr. Lars O. Eriksson for the implementation of the control strategy and for performing the numerical experiments. A comment of Professor C.W. Gear after our talk inspired a substantial change of Section 3.

References

1. Dahlquist, G., Edsberg, L., Sköllermo, G. and Söderlind, G.: Are the Numerical Methods and Software Satisfactory for Chemical Kinetics? Report TRITA-NA-8005, Dept. of Numerical Analysis and Computing Science, Royal Inst. Technology, Stockholm, 1980. To appear in Springer Lecture Notes from a Symposium at Bielefeld, April 1980.

2. Dahlquist, G., Liniger, W. and Nevanlinna, O.: Stability of Two-Step Methods for Variable Integration Steps. IBM Research Report RC 8494, September 1980; submitted to SIAM J. Numer. Anal.

3. Lindberg, B.: Characterizations of Optimal Stepsize Sequences for Methods for Stiff Differential Equations. SIAM J. Numer. Anal., vol.14, 859-887, (1977).

4. Skelboe, S. and Christensen, B.: Backward Differentiation Formulas with Extended Regions of Absolute Stability. BIT, vol.21, 221-231, (1981).

5. Henrici, P.: Discrete Variable Methods in Ordinary Differential Equations. J. Wiley & Sons 1962.

Chebyshev Methods for Integral and Differential Equations

L. M. Delves

1. Introduction

The historical trend in algorithms in all areas of approximation seems to be towards the use of higher order methods. Thus, in quadrature, automatic routines based on the Trapezoidal or Simpsons rule with mesh refinement, have given way to routines based on a sequence of Gauss rules and much larger grids; in ODE's, 4th order RK has given way (maybe?) to variable order Adams methods; and even in the FEM, attention is turning increasingly towards the use of larger "high order" (though so far only modestly high) elements.

However, implementing high-order methods can and does lead to instability or illconditioning problems. These are exemplified by the approximation theory problem:

$$\text{approximate } f(x) \text{ by } f_N(x) = \sum_{i=0}^{N} a_i x^i \text{ on } [0, 1]. \tag{1}$$

In the L_2 norm, this leads to the defining equations

$$A\,\underline{a} = \underline{f} \tag{2}$$

$$f_i = \int_0^1 x^i f(x)\,dx, \qquad i = 0, 1 - - N$$

$$A_{ij} = \int_0^1 x^{i+j}\,dx = \frac{1}{i+j+1}, \; i, j = 0, 1 - - N$$

that is, A is the notoriously ill-conditioned Hilbert matrix.

The solution to the problem of illconditioning in this case is well known; we replace (1) by the orthonormal expansion

$$f_N(x) = \sum_{i=0}^{N} \bar{a}_i \bar{P}_i(x) \tag{1'}$$

where $\bar{P}_i(x)$ is a shifted Legendre polynomial satisfying

$$\int_0^1 \bar{P}_i(x)\,\bar{P}_j(x)\,dx = \delta_{ij}. \tag{3}$$

Then (2) is replaced by the trivially well-conditioned equation

$$I\,\underline{\bar{a}} = \underline{\bar{f}}\ ,\ \bar{f}_i = \int_0^1 \bar{P}_i(x)\ f(x)\ dx \tag{2'}$$

and the ill conditioning problems have vanished. Provided that the coefficients $\bar{f}_i$ can be efficiently computed, the revised algorithm also has speed advantages, since most of the work in (2) goes into solving rather than setting up the equation.

This approximation problem is untypically simple. However, orthogonal expansions form a very natural tool on which to base high order methods; this is especially true if the method is intended to be genuinely variable-order rather than high-but-fixed order. Attempts to base general-purpose algorithms on such expansions have in the past led to code which run extraordinarily slowly on other than model problems; however, we argue here that this was only because of the primitive implementation techniques used, and sketch recent work on methods based on Chebyshev polynomial expansions which yield quite low operation counts. The methods are semi-analytic in nature, being based directly on the analytic manipulation of Chebyshev expansions, with these being provided where necessary by FFT techniques; efficient indirect schemes for both setting up and solving the defining equations are discussed, and a number of numerical examples given to illustrate the techniques.

2. Fredholm Equations of the Second Kind

The simplest illustration of the techniques involved is given by the solution of a linear Fredholm equation of the second kind, of the form

$$f(x) = g(x) + \lambda \int_{-1}^{1} K(x, y)\ f(y)\ dy\ . \tag{4}$$

To emphasize the special role played by Chebyshev polynomial expansions, we consider first a normalised Legendre polynomial expansion of the form

$$f(x) \approx f_N^{(P)}(x) = \sum_{i=0}^{N} a_i^{(P)}\ P_i(x) \tag{5}$$

yielding the Galerkin defining equations

$$\begin{aligned} &[I - \lambda B]\underline{a}^{(P)} = \underline{g}^{(P)} \\ &B_{ij}^{(P)} = \int_{-1}^{1} dx \int_{-1}^{1} dy\ K(x, y)\ P_i(x)\ P_j(y) \qquad i,j = 0,\ 1 \text{--} N \\ &\underline{g}_i^{(P)} = \int_{-1}^{1} dx\ g(x)\ P_i(x)\ . \qquad i = 0,\ 1 \text{ - - } N \end{aligned} \tag{6}$$

The integrals (6) are most naturally evaluated using a Q-point Gauss-Legendre rule with points and weights $\left\{x_k^{\,Q},\ w_k^{\,Q},\ k = 1,\ 2 - Q\right\}$, yielding

$$B_{ij}^{(P)} \approx \sum_{k=1}^{Q} \sum_{l=1}^{Q} P_i(x_k)\ P_j(x_l)\ K(x_k,\ x_l)\ w_k\ w_l \tag{6a}$$

$$g_i = \sum_{k=1}^{Q} g(x_k)\, P_i(x_k)\, w_k. \tag{6b}$$

It is necessary to take $Q \geq N$; then a straightforward implementation of (6a) has operation count $\mathcal{O}(N^4)$, and I have seen programs which "achieve" this cost; but a straightforward splitting of (6a) reduces this to $\mathcal{O}(N^3)$. This is still high; to reduce it further we go back to square one and use a Chebyshev polynomial expansion instead. The Chebyshev polynomials are orthogonal on [-1, 1] with weight factor $(1-x^2)^{-\frac{1}{2}}$:

$$\int_{-1}^{1} \frac{T_i(x)T_j(x)\,dx}{\sqrt{1-x^2}} \begin{array}{ll} = 0 & i \neq j \\ \pi & i = j = 0 \\ \pi/2 & i = j > 0\ . \end{array}$$

We make the expansions

$$f(x) = \sum_{i=0}^{\infty} b_i\, T_i(x) \approx f_N = \sum_{i=0}^{N} a_i\, T_i(x) \tag{7}$$

and obtain the Galerkin equations

$$[D - \lambda\bar{B}]\underline{a} = \underline{\bar{g}}$$

$$D_i = \text{diag}\,(d_{ii});\ d_{00} = \pi,\ d_{ii} = \pi/2,\ i > 0$$

$$\bar{B}_{ij} = \int_{-1}^{1} dx\, \frac{T_i(x)}{\sqrt{1-x^2}} \int_{-1}^{1} dy B(x,y)\, \frac{T_j(y)}{\sqrt{1-y^2}} \tag{8}$$

$$\bar{g}_i = \int_{-1}^{1} dx\, \frac{T_i(x)g(x)}{\sqrt{1-x^2}}$$

where $B(x,y) = K(x,y)\,\sqrt{1-y^2}$.

Equations (8) identify $\bar{g}_i$, $\bar{B}_{ij}$ as coefficients in the expansions

$$\frac{\pi}{2}\, g(x) = {\sum_{i=0}^{\infty}}'\, \bar{g}_i\, T_i(x) \quad ; \quad \frac{\pi^2}{4}\, B(x,y) = {\sum_{i,j=0}^{\infty}}'\, \bar{B}_{ij}\, T_i(x)\, T_j(y)\ . \tag{9}$$

Provided $g(x)$ is smooth, accurate approximations g_i to $\bar{g}_i$ can be obtained using the Gauss-Chebyshev P-point rule:

$$\bar{g}_i \approx g_i = \frac{\pi}{P} {\sum_{k=0}^{P}}''\, g(\cos\frac{k\pi}{P})\, T_i(\cos\frac{k\pi}{P})$$

$$= \frac{\pi}{P} {\sum_{k=0}^{P}}''\, g(\cos\frac{k\pi}{P})\, \cos\,(\frac{ki\pi}{P})\ . \tag{10}$$

and the form of (10) allows g_i, i = 0, 1-P to be accumulated using FFT techniques in $\mathcal{O}$(P ln P) operations; it is shown in [2] that a satisfactory strategy is to set P=N. If (by some miracle) B(x, y) is smooth, the matrix B may be evaluated effectively in $\mathcal{O}(N^2 \ln N)$ operations using a double FFT procedure. However, usually this is not so, and we proceed as follows:

1) If K(xy) is smooth, note that

$$\sqrt{1-y^2} = \frac{2}{\pi}\left[1 - \sum_{l=1}^{\infty} \frac{2}{4l^2-1} T_{2l}(y)\right] \tag{11a}$$

and

(a) Evaluate the coefficients K_{ij} in the expansions

$$\frac{\pi^2}{4} K(x,y) = \sum_{i,j=0}^{\infty}{}' K_{ij} T_i(x) T_j(y) \tag{11b}$$

(b) Evaluate B_{ij} by multiplying the expansions (11a), (11b) together. Both stages (a) and (b) can be carried out using FFT techniques in $\mathcal{O}(N^2 \ln N)$ operations; it is necessary to use 2N terms of (11a). Details and an error analysis are given in [2].

2) If K(x, y) and/or g(x) is <u>not</u> smooth we factor them (if possible) as follows:

$$g = w \times h$$

$$K = W \times Q; \quad H = Q(1-x^2)^{\frac{1}{2}} . \tag{12}$$

We assume that

(a) w, W are smooth: we may expand these as Chebyshev series using FFT
(b) the Chebyshev expansions of h,H are possibly slowly convergent but are known analytically.

Under these conditions, the Galerkin equations can be set up numerically in $\mathcal{O}(N^2 \ln N)$ operations; for details, see [2]. To use this technique effectively, it is assumed that libraries of singular expansions will be built up; some are given in [2], while we have developed others since; an implementation of the scheme is described in [4]. As an example of the effectiveness of the resulting algorithm, we take the following example from Baker [5], who gave a numerical solution based on a low order method using N equally spaced points.

<u>Example 1.</u>

$$f(x) = g(x) + \int_0^1 \ln | x - y | f(y) dy$$

$$g(x) = x - 0.5 \left[x^2 \ln x + (1-x^2) \ln (1-x) - (x + \tfrac{1}{2})\right].$$

Exact solution:

$$f(x) = x$$

Error in Solution of example 1

High order method			Low order method [5]	
N	Error	Error Estimate	N	Error
3	1.7×10^{-2}	7.5×10^{-2}	8	1.0×10^{-3}
5	3.5×10^{-12}	2.1×10^{-11}	16	2.7×10^{-4}
			32	7.1×10^{-5}
			64	1.8×10^{-5}

3. Error Estimates

The table above gives error estimates for the Chebyshev method; these come from the algorithm itself, and it is an advantage of methods based on orthogonal expansions that at least crude error estimates can always be provided cheaply. In the context above, we seek an approximation $f_N = \sum_{i=j}^{N} a_i T_i$ to an exact solution $f = \sum_{i=j}^{\infty} b_i T_i$; the error is trivially bounded as

$$||f - f_N||_\infty \leq \sum_{i=0}^{N} |a_i - b_i| + \sum_{i=N+1}^{\infty} |b_i| = S_1 + S_2$$

where S_1 represents the discretisation error and S_2 the truncation error. The latter can be estimated from the rate of decrease of the computed coefficients a_i, and provided S_1 does not dominate can serve as an indicator of the achieved accuracy. The term S_1 is always more difficult to estimate; however for the second kind Fredholm equations considered above, quite good (and readily computable) estimates of S_1 can be provided in terms of the structure of the Galerkin matrix and right hand side; see [2] for details.

4. First kind Integral Equations

The first kind equation

$$\int_{-1}^{1} K(x,y)\, f(y)\, dy = g(x) \qquad (13)$$

can clearly be formally treated in the same way. Introducing as before the expansion (7), the weighted Galerkin equations for [13] are

$$B\,\underline{a} = \underline{g} \tag{14}$$

where B, $\underline{g}$ are as in (8) and can be computed numerically as before. However, a direct solution of (14) ignores the ill posed nature of (13), and some sort of regularisation procedure is needed to avoid getting numerical nonsense for large N. The procedure suggested in [6] has the aesthetic advantage (well, we think it is pretty) of stemming directly from the orthogonal expansion used. Since we seek a square integrable solution f of (13), the coefficients b_i are certainly $o(i^{-\frac{1}{2}})$ for large i; that is, for some $p > \frac{1}{2}$

$$|b_i| < Ci^{-p} \qquad i > 0 .$$

It is certainly reasonable to impose this condition on the computed coefficients a_i also; we therefore augment (14) with the inequalities

$$|a_i| < C\,i^{-p} \qquad i = 1, 2 - - N, \tag{14'}$$

and solve (14), (14') as an over determined system.

A solution in the ℓ_1 (or ℓ_∞) norm leads to a linear programming problem for which efficient techniques exist; the constants C, p are regularisation parameters which need to be fixed, and suitable techniques for this, together with a proof that the procedure yields a stable (i.e. regularised) algorithm, are given in [6]. We demonstrate both its rapid convergence and its stability with an example:

Example 2

$$\int_0^x \cos(x-y)\, f(y)\, dy = \sin x \quad , \qquad 0 \leqslant x \leqslant 1$$

Exact solution:

$$f(x) = 1$$

Errors in the Augmented Galerkin solution

N	3	5	7	10	15	20
error	$9.3, 10^{-2}$	7.5,-4	1.8,-6	2.9,-9	2.6,-9	6.6,-9

The results show that the error decreases rapidly to about 10^{-8} and then remains constant, with no sign of instability or illconditioning.

Existence of a solution

It is quite easy to pose first kind equations which have no square integrable solution; solutions to (13) only exist if the analytic behaviour of K(x,y) and g(x) are "compatible". The reason for this can be seen heuristically as follows. We recall from (9) that $\underline{g}$, B in (14) contain the expansion coefficients for g(x), B(x,y) respectively; we therefore expect $\underline{g}$ to satisfy a bound of the form (14'), and B to satisfy a similar two dimensional bound:

$$|g_i| \leq C_g i^{-s} \qquad i, j \geq 1$$

$$|B_{ij}| \leq C'_B i^{-p'} j^{-q'} \qquad i > j \tag{15}$$

$$\leq C_B i^{-p} j^{-q} \qquad i \leq j$$

where the reason for the splitting in (15) will become evident below. If we now oversimplify by setting $p = p'$, $q = q'$ and assuming equality in (15), it is easy to see that no solution to the infinite equations (14) exists unless $p = s$. More realistically, we have the following theorem [6].

Theorem

Suppose that in addition to (15), an infinite sequence of integers {Q} exists such that for some $C_g' > 0$

$$|g_i| \geq C'_g i^{-s}, \ i \in \{Q\} \tag{15'}$$

Then a necessary condition for the existence of a solution of the infinite system of equations (14) is

$$s = p' \qquad \text{if} \qquad p' \leq p + q - \tfrac{1}{2} \tag{16a}$$

$$s > p + q - \tfrac{1}{2} \text{ if} \qquad p' > p + q - \tfrac{1}{2} \ . \tag{16b}$$

This theorem provides an existence criterion for any first kind integral equation. The coefficients p, p', q, q', s can be estimated numerically from the computed system (14) (see [6] for details); and then (16) is trivial to check. The augmented Galerkin method thus provides its own "existence proof" as well as a numerical solution; and one which works surprisingly well in practice. Thus, for example (2) we find ([6]) the following estimated parameter values

Parameter values computed for problem 2

N	P'	$p + q - \frac{1}{2}$	s
5	6.26	6.66	5.94
7	8.46	9.27	8.26
10	9.66	9.45	9.66
15	9.65	9.45	9.65
20	8.97	9.46	9.65

For these, it appears that $p' \leqslant p + q - \frac{1}{2}$, and hence that there is no solution unless $s = p'$. However, the estimates for s, p' agree rather well, and certainly there is no case for rejecting this solution.

For an example with no solution, we take
Problem 3

$$\int_{-1}^{1} e^{(x+1)\ (y+1)/4}\ f(y)\ dy = (1-x^2)^{\frac{1}{2}}$$

which cannot have a solution because the right hand side has no finite second derivative at $x^2 = 1$; this singularity cannot be matched on the left side. Computed parameter values for this problen are given below

Parameter values computed for problem 3

N	P'	$p + q - \frac{1}{2}$	s
5	6.52	6.24	2.32
7	8.90	8.77	2.24
10	11.26	8.95	2.17
15	11.07	8.94	2.13
20	8.74	8.94	2.10

From this table, we find that $p' > p + q - \frac{1}{2}$, but $s < p + q - \frac{1}{2}$; that is, we predict numerically and correctly that no solution exists.

5. Differential Equations

We now consider the extension of these techniques to differential equations. For simplicity, we restrict attention to second order ordinary differential equations defined on [- 1, 1], and ignore awkward complications such as general boundary conditions, sub-division of the region and attendant continuity conditions, which must be handled to give a practical method. We consider then the problem

$$\left[- \frac{d}{dx} A(x) \frac{d}{dx} + B(x) \right] f(x) = g(x) \qquad -1 \leqslant x \leqslant 1 \tag{17}$$

subject to suitable boundary conditions.

Method 1

We consider two numerical schemes; the first yields (when elaborated) the Global Element method [7, 8]. Adding boundary conditions to (17), making the expansion

$$f = \sum_{i=0}^{N} a_i \, h_i(x) \tag{18}$$

and minimising a suitable functional (see [7]) yields defining equations of the form

$$[A + B + S] \, \underline{a} = \underline{g} + \underline{H} \tag{19}$$

where

$$A_{ij} = \int_{-1}^{1} h_i' \, A(x) \, h_j' \, dx$$

$$B_{ij} = \int_{-1}^{1} h_i \, B(x) \, h_j \, dx \tag{20}$$

$$g_i = \int_{-1}^{1} h_i \, g(x) \, dx$$

and S, H are boundary condition contributions to the equations which vanish if $f(-1) = f(1) = 0$ and $h_i(-1) = h_i(1) = 0$. We assume the former and ensure the latter by choosing

$$h_i(x) = (1-x^2) \, T_i(x) \, . \tag{21}$$

Then our problem reduces to that of computing the matrices A,B and the vector $\underline{g}$. Now $\underline{g}$ can be computed using the same techniques as for the integral equation problem from a Chebyshev expansion of g(x) or of $g(x)\,(1-x^2)$. We evaluate B and A using similar tricks. Introducing the expansion

$$B(x)\,(1-x^2)^{\alpha} = \sum_{i=0}^{\infty}{}' \, b_i^{(\alpha)} \, T_i(x)$$

and noting that $T_i(x)\,T_j(x) = \frac{1}{2}\left[T_{i+j}(x) + T_{|i-j|}(x)\right]$ we find the identity

$$B_{ij} = \frac{\pi}{4}\left[b_{i+j}^{(5/2)} + b_{|i-j|}^{(5/2)}\right] . \tag{21a}$$

As before, we can find accurate approximations to the coefficients $b_i^{(5/2)}$ by expanding $B(x)\,(1-x^2)^2$ numerically and then multiplying by the expansion of $(1-x^2)^{\frac{1}{2}}$. The cost of this is $\mathcal{O}(N \ln N)$ (for the expansions) + $\mathcal{O}(N^2)$ (to fill the N x N matrix B); this is much lower than a direct evaluation of the N^2 integrals defining the B_{ij}. A similar game, with similar cost, can be played for the matrix A, using the identity

$$\frac{d}{dx}\left[(1-x^2)T_i(x)\right] = \frac{1}{2}\left[\,(i-2)T_{i-1}(x) - (i+2)T_{i+1}(x)\,\right] .$$

See [8] for details.

Method 2

An alternative approach [9] is to solve, not for f but for its highest occuring derivative (in this example for f"). We illustrate the principles involved with the trivial first order equation

$$f'(x) = g(x) \tag{22a}$$

with boundary condition

$$f(b) = e\ . \tag{22b}$$

If we expand $f'(x)$ and $f(x)$

$$f'(x) = \sum_{i=0}^{\infty} a_i'\, T_i(x)$$

$$f(x) = \sum_{i=0}^{\infty} a_i\, T_i(x) \tag{23}$$

then it is well known that these coefficients satisfy the recurrence relations

$$a_j = \frac{1}{2j}\left[a'_{j-1} - a'_{j+1}\right] \qquad j = 1,2, --- \tag{24}$$

which can be written as the infinite matrix relation

$$\underline{a}^{(1)} = A\underline{a}' \tag{24a}$$

when $\underline{a}^{(1)T} = [a_1, a_2 \; - - \qquad]$

and A is banded and triangular. Equation (22) then takes the form

$$\underline{a}' = \underline{g} \tag{25a}$$

$$\underline{a}^{(1)} = A\underline{a}' \tag{25b}$$

$$T^T\underline{a} = e \tag{25c}$$

where

$$T^T = [\tfrac{1}{2}\, T_0(b),\ T_1(b),\ T_2(b),\ - -]$$

and (25) can be solved by truncating the system and solving (25a) first. Applied to the second order system (17), this approach replaces $\underline{a}'$ by $A\underline{a}''$, $\underline{a}$ by $A^2\underline{a}''$, and sets up a truncated system of equations for $\underline{a}''$, taking special care of the first two components of $\underline{a}$ and of the boundary conditions. It has been developed for integro-differential equations in [9], and has the advantage that it uses the same basis (T_i) as the method described for integral equations, and hence extends the scope of this method in a natural way. However, method 1 appears to generalise more readily to partial differential equations; see [2]. Both methods exhibit the rapid convergence (for smooth problems) which we have sought to retain; we illustrate them with two simple examples.

Example 4 (method 1 : Global Element method)

$$\nabla^2 f + \frac{K}{y}\frac{\partial f}{\partial y} = 0\ , \qquad 0 \leqslant x,\ y \leqslant 1$$

$f(x, y) = x - y$ on the edge of the square
solution : not known.

This is ELLPACK problem #29 [10].

Errors in Example 4 using the Global Element Method

N	K = - 1	K = - 3	K = 1
4	7.1,-2	5.1×10^{-2}	8.5,-2
6	1.8,-2	4.3,-3	3.1,-2
8	1.8,-3	1.3,-3	4.6,-3
10	4.7,-4	8.2,-4	1.1,-3

These results are not particularly dramatic; but the problem itself is singular, and quite unpleasant to solve accurately.

Example 5 (method 2 : Fast Galerkin [9])

$$[8\,x^4 - 8\,x^2 + 0.5]\,f'(x) - \lambda \int_{-1}^{1} e^{(x+1)y}\, f(y)\, dy = g(x)$$

$$g(x) = (8x^4 - 8x^2 + 0.5)e^x + \lambda(e^{(x+2)} + e^{-(x+2)})/(x+2)$$

$$f(1) + f(-1) = e + e^{-1}.$$

Solution : $f(x) = e^x$

Errors, and error estimate using method 2 [9]

N	Error, $\lambda = 0$	Error, $\lambda = 1$
4	2.8×10^0	1.2×10^0
6	1.3,-1	6.8,0
8	2.7,-3	6.0,- 3
10	2.1,-5	1.6,- 3
12	2.1,-7	9.7,- 8
14	8.7,-11	4.5,- 9

This example was chosen in [9] to attempt to pick holes in the method; the results are gratifyingly good.

6. Solution of the Defining Equations

Finally, we say a little about solution techniques; here, the aim is again to retrieve an apparently poor situation. The techniques discussed above lead to full matrices both for differential and integral equations; we have shown how to fill them effectively. If we now adopt standard solution techniques, these gains will be lost, with $\mathcal{O}(N^3)$ solution times in one dimension, and $\mathcal{O}(N^6)$ in two dimensions. The obvious remedy is to seek iterative solution methods; and the structure of the equations is such that iterative techniques are in fact available with guaranteed and rapid convergence.

The situation is simplest for Fredholm second kind equations. Here, equation (8) has the form

$$[D - B]\,\underline{a} = \underline{g} \tag{26}$$

where D is diagonal with elements of order unity and the elements B_{ij} satisfy bounds of the form (15). These then imply that only the leading submatrix of B is "large". We therefore partition B in the form

$$\left[\begin{array}{c|c} M\mathrm{x}M & 0 \\ \hline 0 & \diagdown \end{array}\right] + \left[\begin{array}{c|c} 0 & \\ \hline & \delta B \end{array}\right] = B_0 + \delta B$$

and use the iterative scheme

$$\left[\, I - B_0 \,\right] \underline{a}^{(n+1)} = \underline{g} - \delta B\underline{a}^{(n)} \tag{27}$$

$$\underline{a}^{(n+1)} = \underline{a}^{(n)} + \underline{\varepsilon}^{(n+1)}$$

or equivalently

$$\left[\, I - B_0 \,\right] \underline{\varepsilon}^{(n+1)} = \underline{g} - (I-B)\underline{a}^{(n)} = \underline{r}^{(n)} . \tag{28}$$

This scheme can be shown to be rapidly convergent for M "sufficiently large", leading to an $\mathcal{O}(N^2)$ overall solution time. Similar, but more complicated schemes apply to integro-differential and to differential and partial differential equations; the additional complications stem mainly from the boundary conditions in these problems.

In other respects, however, differential equations yield even more favourable iterative schemes than the integral equations. Note that, in (28), the left side involves only a small part (the MxM submatrix) of B; the cost of setting this up is $\mathcal{O}(M^2) << \mathcal{O}(N^2)$. The full matrix B is involved only in order to compute the residual $\underline{r}^{(n)}$; it is possible (see [11]) to compute $r^{(n)}$ directly in $\mathcal{O}(N \ln N)$ operations for an ODE in one dimension ($\mathcal{O}(N^2 \ln N)$ in two space dimensions) and hence to reduce the operations count of both matrix set up and solution phases. For elliptic pde's in two dimensions, the best overall operations count that has been demonstrated is $\mathcal{O}(N^3)$ - but it seems likely that this can be reduced to the $\mathcal{O}(N^2 \ln N)$ needed to form the residual, and these counts are remarkably low for a problem in which we are computing $\mathcal{O}(N^2)$ unknowns.

Summary

Chebyshev polynomial techniques have a number of pleasing features, which the reader may of course rank in any way he/she sees fit. The ranking below is intended to be provocative rather than order-of-importance.

1) Aesthetic. The semi-analytic nature of the techniques leads to rather pretty algorithms; perhaps more satisfying than brute force techniques.

2) Rapid convergence for both smooth and singular problems Provided the singularities of a given problem are well understood, these techniques yield convergence as fast as for smooth problems, and hence should be more economic.

3) Different. They make a change from playing with low order approximations and grid refinement techniques.

On the other side of the picture, the methods do lead to more complex programs than the conventional approach, and attempts to simplify the code by using direct implementation techniques lead to codes with high operation counts and long run times, which in the past have probably been responsible for putting users decidedly off such methods. However, the implementation techniques are now available; codes only have to be written once; and anyway the techniques needed are less familiar rather than more complicated than those used, say, in the FEM game. I think these methods have a future.

References

[1] L. M. Delves, "A Fast Method for the solution of Fredholm Integral Equations" J. Inst. Math. Applics. 20 (1977) 173 - 182.

[2] L. M. Delves, L. F. Abd-Elal & J. A. Hendry "A Fast Galerkin Algorithm for Singular Integral Equations" J. Inst. Math. Applics. 23 (1979) 139-166 .

[3] R. E. Scraton, "A Chebyshev method for the solution of Fredholm Integral Equations" Math. Comput. 23, (1969) 837-845.

[4] L. M. Delves, L. F. Abd-Elal & J. A. Hendry "A set of Modules for the solution of Integral Equations" Comput. Jnl. 24 (1981) 184-190 .

[5] C. T. H. Baker, "The Numerical Treatment of Integral Equations" Oxford University Press 1978.

[6] E. Babolian & L. M. Delves, "An Augmented Galerkin Method for First Kind Fredholm Equations" J. Inst. Math. Applics. 24 (1979) 151-174.

[7] L. M. Delves & C. A. Hall, "An Implicit Matching Principle for Global Element Calculations" J. Inst. Math.Applics. 23(1979) 223-234.

[8] L. M. Delves & C. Phillips, "A Fast Implementation of the Global Element Method" J. Inst. Math. Applics. 25 (1980) 177-197.

[9] E. Babolian & L. M. Delves, "A Fast Galerkin scheme for Linear Integro-differential Equations" IMA Journal of Numer. Anal. 1(1981) 193-213.

[10] E. N. Hasting & J. R. Rice "A Population of Elliptic Partial Differential Equations in two Variables" Preprint, Purdue University, 1978.

[11] L. M. Delves, J. A. Hendry & J. Mohamed, "Iterative solution of the Global Element Equations" submitted to J. Comp. Phys.

Simulation of Miscible Displacement in Porous Media by a Modified Method of Characteristic Procedure

Jim Douglas, Jr.

Abstract. The miscible displacement of one incompressible fluid by another in a porous medium is described by a system of two nonlinear equations , one elliptic in form for the pressure and the other parabolic in form for the concentration. The pressure and the fluid velocity will be approximated by a mixed finite element method, and the concentration by a finite difference method based on the use of a modified method of characteristic procedure. A convergence analysis is given for the method.

1. Introduction.

The miscible displacement of one incompressible fluid by another in a horizontal reservoir $\Omega \times [0,1]$ of unit thickness is described [6,7] by the system

$$\text{(a)} \qquad \nabla \cdot u = -\nabla \cdot \left(\frac{k(x)}{\mu(c)} \nabla p\right) = q \ ,$$

(1.1)

$$\text{(b)} \qquad \phi(x) \frac{\partial c}{\partial t} + u \cdot \nabla c - \nabla \cdot (D\nabla c) = (\tilde{c} - c)q \ ,$$

for $x \in \Omega$ and $t \in J = (0,T]$. The permeability $k(x)$ and the porosity $\phi(x)$ are rock properties; the viscosity $\mu(c)$ depends, usually quite strongly, on the concentration c of one of the fluids. The symbols p and u denote the pressure and the Darcy velocity of the fluid. The diffusion coefficient D is made up of two terms, the first representing molecular diffusion and the second a velocity-dependent, tensorial dispersion [6]; the required brevity of this presentation limits the treatment here to the case of molecular diffusion, $D = D(x) = d_m\phi(x)$. The external flow q will be considered to be smoothly distributed in the analysis below; however, in practical situations it must be considered to take place at sources and sinks. The function $\tilde{c}$ must be specified at points at which fluid is being injected into the reservoir $(q > 0)$ and will be assumed equal to c at points at which fluid is being produced $(q < 0)$. The interest in miscible displacement for engineering purposes is primarily in the interior behavior of the flooding process; in order to center the attention below on the interior, the domain Ω will be assumed to be a square and the problem will be assumed periodic with period Ω. Thus, no boundary conditions need be specified. The initial values of the concentration must be given; those for the pressure can then be calculated. Note that the pressure is determined up to an additive constant, so long as the net flow, $(q,1)$, is zero at all times.

The concentration equation (1.1.b) requires the Darcy velocity u from the pressure equation, but not the pressure itself. Thus, it is appropriate to choose a numerical method for the pressure equation (1.1.a) that approximates the velocity directly, and a mixed finite element method will be selected. Physically, the convection dominates the diffusion in the flow; consequently, it is helpful to apply a numerical method that is intended to follow the flow. This can be done by treating the transport $\phi c_t + u \cdot \nabla c$ by a method of characteristic technique. In order to do this, first transform (1.1.b) into the form

$$\psi(x,t,u) \frac{\partial c}{\partial \tau} - \nabla \cdot (D\nabla c) = (\tilde{c} - c)q \ , \tag{1.2}$$

where

(1.3)

(a) $$\psi(x,t,u) = \{\phi(x)^2 + |u(x,t)|^2\}^{1/2} \ ,$$

(b) $$\frac{\partial}{\partial \tau} = \psi^{-1}(\phi \frac{\partial}{\partial t} + u \cdot \nabla) \ .$$

A finite difference procedure will be defined to discretize (1.2), with the time direction being replaced locally by the characteristic $\tau = \tau(x,u)$.

The analysis of the combined finite element-finite difference procedure will depend strongly on the author's treatment [3] of a finite difference method for miscible displacement and the joint work [4] of the author, R.E. Ewing, and M.F. Wheeler on a finite element approach that used the mixed finite element method for the pressure.

2. Discretization of the Pressure Equation. Let h_c define the (uniform) grid spacing for the concentration and let $C^n_{ij} = C^n(x_{ij}) = C(t^n, ih_c, jh_c)$ denote the approximation to the concentration at (t^n, x_{ij}). Extend the mesh function to $C^n(x)$ by piecewise bilinear interpolation. Cover Ω by a rectangular, quasi-regular quadrilateralization with associated discretization parameter h_p. Let $V_{h_p} \times W_{h_p}$ be the periodic Raviart-Thomas space [8] of index k (i.e., permitting approximations to $O(h_p^{k+1})$) contained in $H(\text{div}; \Omega, \text{periodic}) \times L^2(\Omega, \text{periodic})$ over the quadrilateralization. Set

$$A(c; u, v) = (\frac{\mu}{k(c)} u, v) = \sum_{i=1}^{2} (\frac{\mu}{k} u_i, v_i) \tag{2.1}$$

for $u, v \in V = H(\text{div}; \Omega, \text{periodic})$ and

$$B(u,\gamma) = -(\nabla \cdot u, \gamma) \tag{2.2}$$

for $u \in V$ and $\gamma \in W = L^2(\Omega$, periodic). The inner products are in $L^2(\Omega)$ or $L^2(\Omega)^2$, as appropriate.

The mixed finite element method for the pressure consists of finding $U^n \in V_{h_p}$ and $P^n \in W_{h_p}$ satisfying the saddle point problem [2,4]

$$(2.3)\quad \begin{aligned} &\text{(a)}\quad A(C^n;U^n,v) + B(v,P^n) = 0 \;, && v \in V_{h_p} \;, \\ &\text{(b)}\quad B(U^n,\gamma) = -(q,\gamma) \;, && \gamma \in W_{h_p} \;. \end{aligned}$$

If $\xi^n = c^n - C^n$ and if $\|\cdot\|_j$ denotes the norm in the periodic Sobolev space $H^j(\Omega$, periodic) and $|\cdot|_j$ denotes the norm in the corresponding discrete space over the concentration mesh, it is shown in [4] that, with $M(\cdot,\cdot,\ldots)$ denoting a generic function of the arguments,

$$(2.4)\quad \begin{aligned} \|u^n - U^n\|_V + \|p^n - P^n\|_W &\le M(\|p^n\|_{k+3})(h_p^{k+1} + \|\xi^n\|_0) \\ &\le M(\|p^n\|_{k+3}, \|c^n\|_2)(h_p^{k+1} + h_c^2 + |\xi^n|_0) \;. \end{aligned}$$

3. Discretization of the Concentration Equation. Let U_{ij}^n denote the evaluation of U^n at x_{ij}. Then, set

$$(3.1)\quad \begin{aligned} &\text{(a)}\quad \overline{X}_{ij}^n = x_{ij} - U_{ij}^n \Delta t/\phi_{ij} \;, \quad \overline{C}_{ij}^n = C^n(\overline{X}_{ij}^n) \;, \\ &\text{(b)}\quad \overline{x}_{ij}^n = x_{ij} - u_{ij}^{n+1} \Delta t/\phi_{ij} \;, \quad \overline{c}_{ij}^n = c(t^n,\overline{x}_{ij}^n) \;. \end{aligned}$$

It is easy to see that

$$(3.2)\quad \left(\psi \frac{\partial c}{\partial \tau}\right)(t^{n+1},x_{ij}) = \phi_{ij} \frac{c_{ij}^{n+1} - \overline{c}_{ij}^n}{\Delta t} + O\left(\left|\frac{\partial^2 c}{\partial \tau^2}\right| \delta\tau\right) ,$$

where the $\partial^2 c/\partial \tau^2$ is evaluated somewhere on the segment between (t^{n+1},x_{ij}) and $(t^n,\overline{x}_{ij}^n)$. Thus, it is natural to approximate (1.2) by the difference equation (for a periodic C^{n+1})

$$(3.3)\quad \phi_{ij} \frac{C_{ij}^{n+1} - \overline{C}_{ij}^n}{\Delta t} - \nabla_h(D\nabla_h C_{ij}^{n+1}) = (\tilde{C}_{ij}^{n+1} - C_{ij}^{n+1})q \;,$$

where the diffusion term is a standard five-point difference and $\tilde{C}_{ij}^{n+1} = \tilde{c}_{ij}^{n+1}$ if $q_{ij} > 0$ and $\tilde{C}_{ij}^{n+1} = C_{ij}^{n+1}$ if $q_{ij} < 0$. Note that C^0 is known from the initial data. Then, if C^n is known, (2.3) can be used to find U^n and P^n and then (3.1.a) and (3.3) to find C^{n+1}. Thus, the computational procedure is completely defined, except, of course, for methods for solving of the linear algebraic system, a problem not to be considered here.

4. The Convergence Analysis. Note that $\tilde{\xi}_{ij}^{n+1} = 0$ if $q_{ij} > 0$ and $\tilde{\xi}_{ij}^{n+1} = \xi_{ij}^{n+1}$ if $q_{ij} < 0$. Denote the positive part of q by $q^+ = \max(q,0)$. Since only molecular diffusion is being considered here, the error equation [3, (4.15)] of [3] simplifies to read

$$\phi_{ij}\frac{\xi_{ij}^{n+1} - (c^n(\bar{x}_{ij}^n) - \bar{C}_{ij}^n)}{\Delta t} - \nabla_h(D\nabla_h\xi_{ij}^{n+1}) + q_{ij}^+\xi_{ij}^{n+1} = \varepsilon_{ij}^n , \tag{4.1}$$

where, if $\|\cdot\|_{j,r}$ denotes the norm in $L^\infty(0,T;W^{j,r}(\Omega))$,

$$|\varepsilon_{ij}^n| \leq M(\|c\|_{3,\infty}, \|\frac{\partial^2 c}{\partial\tau^2}\|_{0,\infty})(h_c + \Delta t). \tag{4.2}$$

Extend the knot values $\{\xi_{ij}^n\}$ to a function $\xi^n = I_1\{\xi_{ij}^n\}$ by piecewise bilinear interpolation, and write $(\bar{\xi}_{ij}^n = \xi^n(\bar{X}_{ij}^n))$

$$\begin{aligned}\xi_{ij}^{n+1} - (c^n(\bar{x}_{ij}^n) - \bar{C}_{ij}^n) &= (\xi_{ij}^{n+1} - \bar{\xi}_{ij}^n) - (c^n(\bar{x}_{ij}^n) - c^n(\bar{X}_{ij}^n)) \\ &\quad + ((1 - I_1)c^n(\bar{X}_{ij}^n) .\end{aligned} \tag{4.3}$$

First,

$$|((1 - I_1)c^n(\bar{X}_{ij}^n)| \leq M\|c\|_{2,\infty}\min(h_c^2, h_c\Delta t) , \tag{4.4}$$

with the $h_c\Delta t$-term arising when $|U_{ij}^n|\Delta t\phi_{ij}^{-1} < \frac{1}{2}h_c$. Next,

$$\begin{aligned}|c^n(\bar{x}_{ij}^n) - c^n(\bar{X}_{ij}^n)| &\leq \|c\|_{1,\infty}|\bar{x}_{ij}^n - \bar{X}_{ij}^n| \leq M\|c\|_{1,\infty}|u_{ij}^{n+1} - U_{ij}^n|\Delta t \\ &\leq M\|c\|_{1,\infty}(|u_{ij}^n - U_{ij}^n| + |u_{ij}^{n+1} - u_{ij}^n|)\Delta t .\end{aligned} \tag{4.4}$$

Hence,

$$\phi\frac{\xi_{ij}^{n+1} - \bar{\xi}_{ij}^n}{\Delta t} - \nabla_h(D\nabla_h\xi_{ij}^{n+1}) + q_{ij}^+\,\xi_{ij}^{n+1} = \tilde{\varepsilon}_{ij}^n , \tag{4.5}$$

where

$$\begin{aligned}|\tilde{\varepsilon}_{ij}^n| &\leq M_1(\|c\|_{3,\infty}, \|\frac{\partial^2 c}{\partial\tau^2}\|_{0,\infty}, \|\frac{\partial u}{\partial t}\|_{0,\infty})(h_c + \Delta t) \\ &\quad + M_2(\|c\|_{1,\infty})|u_{ij}^n - U_{ij}^n| .\end{aligned} \tag{4.6}$$

The appearance of a norm of $\partial u/\partial t$ is not discouraging, as the velocity field changes much less rapidly with time than either the pressure or the concentration.

An $\ell^2(\Omega)$-estimate of ξ is derivable using ξ^{n+1} as the test function for (4.5). It follows from (2.4) and (4.5) that

$$\frac{1}{\Delta t}\{<\phi\xi^{n+1},\xi^{n+1}> - <\phi\bar{\xi}^n,\xi^{n+1}>\} + <D\nabla_h\xi^{n+1},\nabla_h\xi^{n+1}> + <q^+\xi^{n+1},\xi^{n+1}>$$

$$\leq M_3(\|c\|_{3,\infty},\|\frac{\partial^2 c}{\partial\tau^2}\|_{0,\infty},\|p\|_{k+3,2},\|\frac{\partial u}{\partial t}\|_{0,\infty})(|\xi^n|_0^2 + |\xi^{n+1}|_0^2 + h_p^{2k+2} + h_c^2 + (\Delta t)^2) + M(\|p\|_{3,2})h_p^4 \quad , \tag{4.7}$$

where the last term comes from replacing the $L^2(\Omega)$-norm of $u - U$ by the $\ell^2(\Omega)$-norm. One consequence of this term is that there is little point in choosing the index k of the Raviart-Thomas space to exceed one; consequently, assume that

$$k = 0 \text{ or } 1 \; . \tag{4.8}$$

Under (4.8), the last term in (4.7) can be dropped. The next step of the argument is to relate $\bar{\xi}^n$ to ξ^n and some smaller terms in the inner product $<\phi\bar{\xi}^n,\xi^{n+1}>$. The development given in [3, (4.26)-(4.40)] can be applied with only a modest modification to the current procedure. Write $\bar{\xi}^n_{ij}$ in the form

$$\begin{aligned}\bar{\xi}^n_{ij} &= \xi^n(x_{ij} - u^n_{ij}\Delta t/\phi_{ij}) + [\xi^n(x_{ij} - U^n_{ij}\Delta t/\phi_{ij}) - \xi^n(x_{ij} - u^n_{ij}\Delta t/\phi_{ij})] \\ &= \xi^n(x_{ij} - u^n_{ij}\Delta t/\phi_{ij}) + \nu^n_{ij} \; .\end{aligned} \tag{4.9}$$

From [3, (4.31)],

$$\sum_{ij}\phi_{ij}\xi^n(x_{ij} - u^n_{ij}\Delta t/\phi_{ij})^2 h_c^2 \leq [1 + M(\|c\|_{1,\infty},\|p\|_{2,\infty})\Delta t] <\phi\xi^n,\xi^n> \quad . \tag{4.10}$$

Also, it follows from [3, (4.38)] and the embedding theorem of Bramble [1] quoted in [3, (4.34)] that

$$\begin{aligned}|<\phi\nu^n,\xi^{n+1}>| &\leq M_3\Delta t <|u^n - U^n||\nabla_h\xi^n|,|\xi^{n+1}|> \\ &\leq M_3\Delta t|\xi^{n+1}|_{\ell^\infty}|u^n - U^n|_0|\nabla_h\xi^n|_0 \\ &\leq M_3\Delta t(\log\frac{1}{h_c})^{1/2}|u^n - U^n|_0|\xi^{n+1}|_1|\nabla_h\xi^n|_0 \; .\end{aligned} \tag{4.11}$$

The relations (4.10) and (4.11) can be utilized in (4.7) to obtain the inequality

$$\frac{1}{2\Delta t}\{\langle\phi\xi^{n+1},\xi^{n+1}\rangle - \langle\phi\xi^{n},\xi^{n}\rangle\}$$

(4.12) $$+ d_*|\nabla_h\xi^{n+1}|_0^2 - M_4(\log\frac{1}{h_c})^{1/2}|u^n - U^n|_0\{|\nabla_h\xi^{n+1}|_0^2 + |\nabla_h\xi^n|_0^2\}$$

$$\leq M_3\{(1 + M_4(\log\frac{1}{h_c})^{1/2}|u^n - U^n|_0)|\xi^{n+1}|_0^2 + |\xi^n|_0^2 + h_p^{2k+2} + h_c^2 + (\Delta t)^2\} ,$$

where d_* is a lower bound for $d_m\phi(x)$ and

(4.13) $$M_4 = M(M_3,\|p\|_{2,\infty}) .$$

Make the induction hypothesis that

(4.14) $$\max_n |u^n - U^n|_0(\log\frac{1}{h_c})^{1/2} \to 0$$

as Δt and $h = (h_c,h_p)$ tend to zero. Since (4.14) implies that $d_* - M_4|u^n - U^n|_0(\log h_c^{-1})^{1/2} > \frac{1}{2}d_*$ for h and Δt sufficiently small and that the coefficient of $|\xi^{n+1}|_0^2$ on the right-hand side is bounded, it follows from (4.12) and the trivial inequality $|\xi^0|_1 \leq M_4 h_c$ that

(4.15) $$\max_n|\xi^n|_0 + (\sum_{j=0}^{T/\Delta t}|\nabla_h\xi^j|_0^2\Delta t)^{1/2} \leq M_4\{h_p^{k+1} + h_c + \Delta t\} .$$

It follows from (2.4) and (4.15) that

(4.16) $$|u^n - U^n|_0 \leq M_4\{h_p^{k+1} + h_c + \Delta t\} ,$$

so that, if

(4.17) $$(h_p^{k+1} + \Delta t)(\log\frac{1}{h_c})^{1/2} \to 0$$

as h and Δt tend to zero, the induction hypothesis will hold and the estimates (4.15) and (4.16) have been established. It also follows from (2.4) and (4.15) that

(4.18) $$\max_n[\|u^n - U^n\|_V + \|p^n - P^n\|_W \leq M_4\{h_p^{k+1} + h_c + \Delta t\} .$$

5. Extensions and Remarks. A number of extensions of the above convergence results are apparent. First, the diffusion coefficient D can take the more general, tensorial form, as can be seen by considering the way the argument of [3] was applied here. Next, since only first-order convergence was obtained with respect to h_c, a non-uniform grid could have been assumed for the concentration. Third, it is usual

to employ longer time steps for the pressure than for the concentration; again a study of [5] will convince the reader that such a choice could have been analyzed for this method. Fourth, the Raviart-Thomas space based on triangles instead of rectangles could have been chosen, with the index still limited to zero or one.

It is important to note that the meshes for pressure and concentration are not required to be be related. In fact, for the $k = 1$ space h_p would be taken to be about the square root of h_c.

References

1. J.H. Bramble, A second-order finite difference analog of the first biharmonic boundary value problem, Numer. Math. 4(1966), pp.236-249.
2. F. Brezzi, On the existence, uniqueness and approximation of saddle-point problems arising from Lagrangian multipliers, RAIRO, Anal. Numér., 2(1974), pp. 129-151.
3. J. Douglas, Jr., Finite difference methods for two-phase, incompressible flow in porous media, to appear.
4. _____, R.E. Ewing, and M.F. Wheeler, The approximation of the pressure by a mixed method in the simulation of miscible displacement, to appear.
5. _____, _____, and _____, A time-stepping method for a miscible displacement simulation technique, to appear.
6. D.W. Peaceman, Improved treatment of dispersion in numerical calculation of multidimensional miscible displacement, Soc. Pet. Eng. J. (1966), pp. 213-216.
7. _____, <u>Fundamentals of Numerical Reservoir Simulation</u>, Elsevier Publishing Co., 1977.
8. P.A. Raviart and J.M. Thomas, A mixed finite element method for 2nd order elliptic problems, <u>Mathematical Aspects of the Finite Element Method</u>, Lecture Notes in Mathematics 606, Springer, 1977.

FULL MATRIX TECHNIQUES IN SPARSE GAUSSIAN ELIMINATION

Iain S. Duff

Abstract

We discuss ways in which code for Gaussian elimination on full systems can be used in crucial parts of the code for the solution of sparse linear equations. We indicate the benefits of using full matrix techniques in the later stages of Gaussian elimination and describe frontal and multi-frontal schemes where such benefits are obtained automatically. We also illustrate the advantages of such approaches when running sparse codes on vector machines.

1. Introduction

Special codes for the direct solution of sets of linear equations whose coefficient matrix is large and sparse have been developed and used for about twenty years. These codes utilise sometimes quite complex data structures to avoid storing or operating on all or most of the zero entries present in the original system. Throughout this paper, we will use the term sparse code to mean this type of code, and the term full code to mean that which treats the matrix as a two-dimensional array and takes little or no account of the presence of zeros. The solution to many problems has only been made tractable by the use of sparse codes. We illustrate the gains of using sparse code over full code by the results shown in Table 1.1. The test examples used in this table and in the others in this paper are described by Duff and Reid (1979). In this case and in the other experimental results presented later, the runs shown are chosen to be representative of many more runs on other test cases.

The sparse code used in these runs (and in those in Tables 1.2 and 1.3) is MA28, the solver for general sparse unsymmetric systems in the Harwell Subroutine Library (Duff (1977)). The full codes are the NAG routines F01BTF and F04AYF (for factorization and solution respectively) (Du Croz et al (1978)).

If we look at the times in Table 1.1, we see that in all but the smallest cases the sparse code is significantly faster than the full code particularly if we are solving a system for which a suitable pivotal sequence is already known. This could be the case if we are solving several problems with the same sparsity structure in, for example, the solution of a non-linear problem. Full codes do not have a similar facility since the work in choosing pivots is an order of magnitude less than that for the decomposition. In such instances, our savings in time by using sparse code is two orders of magnitude, significant by any yardstick. Of course, the limitation on storage is a prime reason why it is impossible to use full codes on very large problems since, for a system of order n, the n^2 words required soon exceed the

		147	292	363	541	199
Order n Non-zeros τ		147 2449	292 2208	363 3068	541 4285	199 701
Times for factorization (pivot order unknown)	SPARSE	.859	.534	.271	1.252	.117
	FULL	1.313	10.002	19.011	62.443	3.248
Times for factorization (pivot order known)	SPARSE	.198	.137	.105	.335	.031
Time for solution	SPARSE	.011	.012	.011	.027	.005
	FULL	.025	.097	.151	.332	.046
Space required						
SPARSE: During factorization	I*	9877	10739	10322	23183	4214
	R*	5803	5611	3624	12610	1523
SPARSE: For solution	I*	6307	6713	5023	14701	2279
	R*	5866	5837	3934	13078	1682
FULL		21609	85264	131769	292681	39601

Table 1.1 Comparison of sparse and full code. Times in seconds on an IBM 3033.

*I: integer storage in words R: real storage in words

capacity of most machines. We show comparable figures for a sparse code in our table, where we have chosen to show integer and real storage separately since, on some machines, the storage for these two data types can differ by a factor of four. On the largest case in our table (of order 541), we see over an order of magnitude saving in storage. Indeed, it is at larger orders still where sparse codes demonstrate their true powers.

The storage requirement for full codes is n^2 words while that for sparse codes depends not only on the order, n, but also on the number of non-zeros, τ, and can vary from a small multiple of τ and n to a small multiple of $\tau \log n$ where the structure militates against the most efficient use of sparse Gaussian elimination. The asymptotic behaviour of the factorisation times even more favours sparse codes where the $O(n^3)$ performance of full codes must be compared with times for sparse codes which typically vary as τ'^2/n, where τ' is the number of non-zeros in the LU factors. For sparse codes on very sparse systems, the number of multiplications can be of order n. Indeed, it is not possible in general to forecast this easily since, as the results of Table 1.1 illustrate, the times are not simply related to order and density and structure plays a very important part in their determination. However, whereas sparse codes often almost routinely solve problems of order 10,000 or more, their solution by full code would require 100 Megawords and approximately 90 hours on an IBM 3033.

Researchers in sparse matrices have, of course, been telling you all this for decades so hopefully this affirmation of the necessity and value of such sparse techniques is not news to you. However, we feel it is most important to emphasise and catalogue these gains in this introduction because we wish to stress the near heresy involved in the title of this paper, namely the use of full codes in sparse Gaussian elimination.

Why might we wish to use full codes in such instances given the aforementioned weight of evidence stacked against them? Certainly such codes are very much simpler and shorter than their sparse counterparts, and additionally the results of Table 1.1 indicate that full codes become competitive at the lower orders and when the matrix is denser. We show in Table 1.2, the effect of changing the density of a sparse matrix of fixed order and see that the break-even point between sparse and full codes is at quite a low density.

Density	.04	.06	.08	.1	.3	Times for full code
Pivot selection and factorisation	.119	.225	.399	.496	1.605	.430
Factorisation	.029	.054	.093	.124	.412	-
Solution	.005	.004	.006	.006	.012	.012

Table 1.2 Times for general sparse code (MA28) on random matrices of order 100 with different densities

If we look at Figure 1.1, which shows the pattern of the filled in matrix when pivoting down the diagonal in order, we see that, towards the end of the elimination the reduced matrix satisfies these two conditions, that is, it is both dense and relatively small. Thus it is evident that it may pay us to use full code towards the end of Gaussian elimination. We discuss this further in section 2.

However, if we order the rows and columns of the reduced matrix so that the non-zeros in the pivot row are in the first columns of the block and the non-zeros in the pivot column are in the first rows, we will have a submatrix of the form

```
Pivot↘
      * x x x
      x
      x
      x
      x
```

which is clearly, after the elimination step corresponding to this pivot, a full submatrix. Indeed, it is by making use of this observation that the very fast ordering techniques for symmetric systems have been developed by George et al (1980),

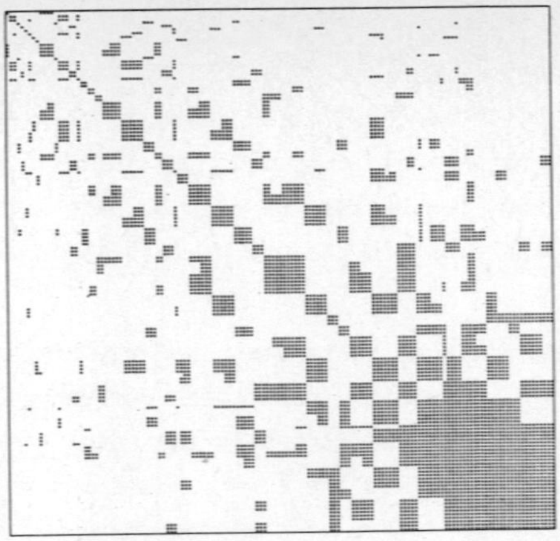

Figure 1.1 Filled-in pattern of matrix of order 147 after factorization using minimum degree ordering

Eisenstat et al (1977) and Duff and Reid (1981a) (see, for example, the book by George and Liu (1981)). Again, it would seem sensible to use full code to take advantage of this although the organisational problems (for example, the permutations implicitly alluded to above) are immense. It is only very recently that techniques have been developed to capitalise on this situation. We discuss these in section 3.

There are an increasing number of machines available which can perform operations on vectors very efficiently, the CRAY-1 being one example. The indirect addressing necessary in the innermost loop of a general sparse code militates against such vectorization whereas the simpler loop of full code vectorizes easily. This is indicated dramatically in Table 1.3 which shows the same codes run on the same test examples as for Table 1.1 but this time on a CRAY-1 computer.

Order Non zeros		147 2449	292 2208	363 3068	541 4285	199 701
Times for factorization (pivot order unknown)	SPARSE	.475	.285	.136	.647	.065
	FULL	.129	.739	1.318	3.907	.274
Times for factorization (pivot order known)	SPARSE	.085	.059	.049	.139	.014
Time for solution	SPARSE	.005	.005	.005	.012	.002
	FULL	.002	.007	.010	.020	.003

Table 1.3 A comparison of full and sparse codes on the CRAY-1. Times in seconds on the CRAY-1.

Although the $O(n^3)$ complexity of full code will still eventually dominate, the break even point of problem size has swung dramatically in favour of full codes. We would expect this swing to be evident in any sparse techniques which use full code and we illustrate this in section 4.

2. Use of full code towards the end of Gaussian elimination

It is unarguable that we can benefit from switching to full code when the reduced matrix is full. It is also clear, for example from the results of Table 1.2, that substantial gains should be obtained by switching when the reduced matrix is still quite sparse.

The Harwell Subroutine Library code MA31 (Munksgaard (1980)) which solves symmetric positive definite systems has an option for switching to full code when the reduced matrix reaches a user set density and we have run that on our test examples and show a sample of the results in Table 2.1.

Density of active matrix when switch to full code occurs (in percentages)	Order / Non zeros	147 / 1298	1176 / 9864	292 / 1250	130 / 713
10	Time for factorization	.676	4.819	.494	.216
20		.269	3.957	.261	.093
30		.196	2.143	.206	.076
40		.183	1.460	.188	.071
50		.183	1.209	.186	.070
60		.189	1.184	.185	.069
70		.195	1.134	.188	.068
80		.204	1.105	.190	.068
90		.214	1.143	.192	.069
100		.228	1.307	.193	.069
no switch		.262	2.053	.203	.077

Table 2.1 Effect of switching to full code towards the end of Gaussian elimination

For most sparse codes, it is relatively easy to switch to using full code at the end of the elimination. For example, if the sparse data structure holds the matrix non-zeros by rows all that is required before using full code is that the columns in each row must be in the same sequence (for some sparse codes this will be the case already) and if the matrix is not full the data structure must be expanded to incorporate all zero entries explicitly. It is particularly simple in the positive definite case because no further pivoting is required and the ordering of the remaining reduced matrix will be unaltered.

However, although this switch can be made very easily, it is not, to our knowledge, incorporated in many sparse codes. Apart from MA31, the only other code we know that uses it is the DMOOP routine of the IBM package SLMATH (IBM (1976)) based on work by Fred Gustavson. Here the switch over density is set within the code and full code is only used when the reduced matrix is full. It is planned to incorporate such a change in a future release of our MA28 package.

Indeed, it is not only at the end of Gaussian elimination that dense blocks occur. For some problems, for example in the power network from a set of test matrices supplied by John Lewis (Boeing Computer Services) dense blocks occur in the formulation of the problem itself. We show this pattern in Figure 2.1.

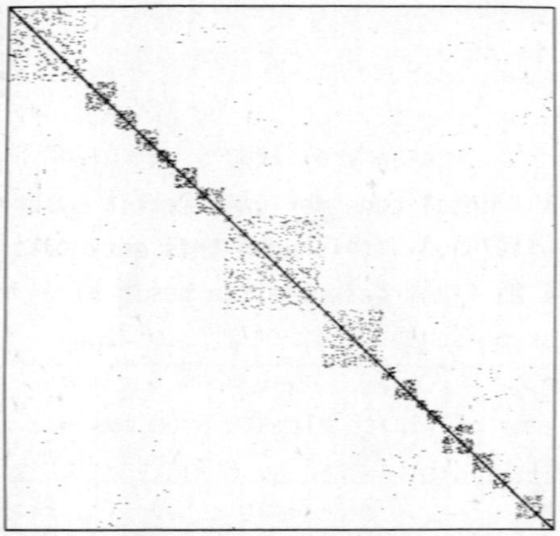

Figure 2.1 Matrix representing Western US power network

In other cases, particularly with orderings other than minimum degree, dense blocks occur at intermediate stages of Gaussian elimination. We illustrate this in Figure 2.2, where we show the filled in pattern after ordering using the nested dissection ordering, a very efficient, popular and easily calculated ordering for grid based problems.

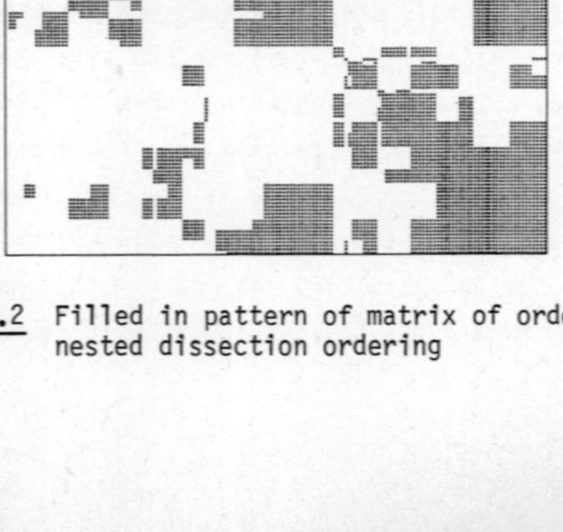

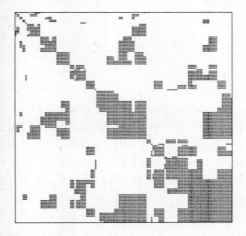

Figure 2.2 Filled in pattern of matrix of order 147 after factorization using nested dissection ordering

To cater for these intermediate dense blocks, we would like an algorithm which automatically detects them and switches to full code on their detection. In the next section we discuss methods which effectively do this.

Another benefit of switching to full code is that the resulting code will, as we witnessed in Table 1.3, vectorize easily. We examine this aspect further in section 4.

3. Frontal and multi-frontal schemes

A class of schemes which utilize full matrix code at the inner loop of sparse Gaussian elimination are the frontal schemes used for many years in the solution of problems in structural analysis. In 1970, Irons published a code for symmetric positive definite systems and a frontal code for unsymmetric systems by Hood (1976) was improved by Cliffe et al (1978). It is upon this work that our code MA32 (Duff (1981a,1981b)) is based. We first describe the basis of a frontal scheme indicating how the computation is organised so that all inner loop calculations are performed using full code.

We will discuss frontal schemes in terms of finite element problems although our frontal code (MA32 Duff (1981a)) can be used to solve any set of unsymmetric linear equations. In a finite element problem the matrix A is a sum

$$A = \sum_{\ell} B^{(\ell)} \tag{3.1}$$

where each $B^{(\ell)}$ has non-zeros in only a few rows and columns and corresponds to contributions to the matrix from finite element ℓ. It is normal to hold $B^{(\ell)}$ in packed form as a small full matrix together with an indexing vector to identify where the non-zeros belong in A. The basic "assembly" operation when forming A is thus of the form

$$a_{ij} \leftarrow a_{ij} + b_{ij}^{(\ell)} \tag{3.2}$$

and it is evident that, if we examine the basic operation in Gaussian elimination viz.

$$a_{ij} \leftarrow a_{ij} - a_{ik} [a_{kk}]^{-1} a_{kj} \tag{3.3}$$

the elimination step (3.3) may be performed before the assembly step (3.2) so long as the terms in the triple product in (3.3) are all fully-summed (that is, have no more sums of the form (3.2) to come) before execution of the elimination operation (3.3). This observation is used in frontal codes by judiciously choosing an assembly and elimination order so that eliminations are performed on relatively small partially assembled matrices. For example, in Figure 3.1, if we assemble the elements in the order indicated by the numbering in that figure, then, in the absence of numerical pivoting, the partially assembled matrix (which we call a frontal matrix) need never exceed the order of the number of variables in two elements.

1	3	5	7	9	11
2	4	6	8	10	12

Figure 3.1 Assembly order in a finite element problem

Indeed, if we permute all the fully-summed variables in the frontal matrix to the first rows and columns, it will have the form shown in Figure 3.2. Pivots can be chosen from anywhere within the doubly-shaded region and, if the assembled

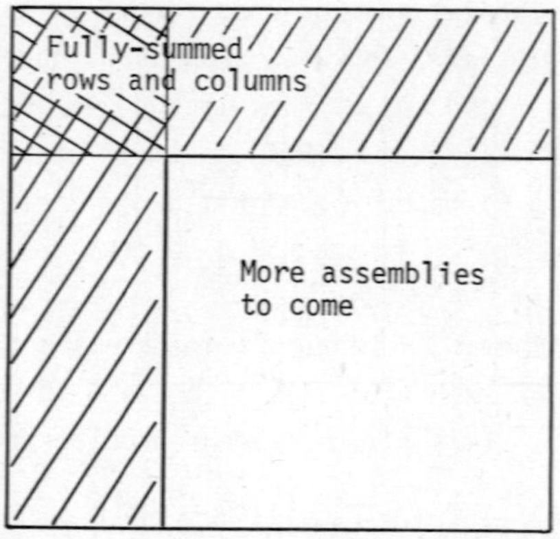

Figure 3.2 Frontal matrix

matrix is symmetric positive definite, could be chosen down the diagonal in order. In the non positive definite case, a numerical pivoting strategy can be employed which chooses diagonal pivots of order 1 or 2 or entries from within the doubly-shaded block from off the diagonal which satisfy some numerical criterion. We can use such strategies (see Duff (1981a), for example) to perform stable decompositions when the matrix is indefinite or unsymmetric at the cost of possibly increasing the order of the frontal matrix slightly. However, for the purposes of our present discussion, the crucial feature is that the frontal matrix of Figure 3.2 is held as a full matrix and all Gaussian elimination operations (3.3) are performed within such matrices using full matrix code.

For non-element problems, we "assemble" the rows (equations) one at a time and a variable becomes fully-summed whenever there are no further equations in which it appears.

We illustrate the performance of our frontal code MA32 on a model problem by comparing it with a general solver for unsymmetric matrices (MA28, Duff (1977)) on 5-point discretizations of the Laplacian operator on rectangular grids. We show these results in Table 3.1. We see that the core requirements of the frontal code are much less than for the general code and its execution time is quite competitive with MA28-FACTOR and much better than the MA28-ANALYZE/FACTOR.

Grid m n	10 10	10 40	10 60	10 100	10 300	32 32	64 64
Decomposition time							
MA32	60	240	370	610	1800	2000	23000
MA28*	100	560	850	1410	5000	8700	N.A.
MA28**	30	130	210	350	1100	1100	N.A.
Solution time							
MA32	10	45	60	100	310	250	1900
MA28	5	15	20	40	120	60	300+
Storage in kbytes							
MA32	20	20	20	20	20	50	150
MA28	15	70	110	180	530	320	1300+

Table 3.1 Times (in msecs on an IBM 3033) on a model problem on an mxn grid

+Estimated

*Time for pivot selection and factorization (MA28-ANALYZE/FACTOR)

**Time for factorization after pivot sequence is known (MA28-FACTOR)

Several users of MA32 have reported runs on large finite element problems where the total run time only exceeds the basic machine time for the arithmetic operations by less than 20%, in spite of storing the matrix factorization out-of-core. Some large finite difference problems on irregular meshes which could not be handled by MA28 or iterative methods are now being solved at Harwell using MA32.

However, the simple frontal scheme described above relies on a geometry which allows the variables to be numbered so that the front size is small relative to the overall size of the problem. For example, if we number the variables in limb 1 of the cross-shaped region shown in Figure 3.4 from the left-hand end all will be well until we reach the situation where our front is indicated by the double line at the rightmost end of that limb. At this stage, if we continue with a single front our frontwidth will increase considerably and the efficiency of the method will degrade substantially. However, the frontal matrix at that stage consists of equations connecting the non-fully summed variables of the double line and indeed is essentially the same as one of the $B^{(\ell)}$ in equation (3.1). We can therefore remove this "generated element" from our frontal matrix storage and continue assembling and eliminating elsewhere in our structure, say on limb 2, starting

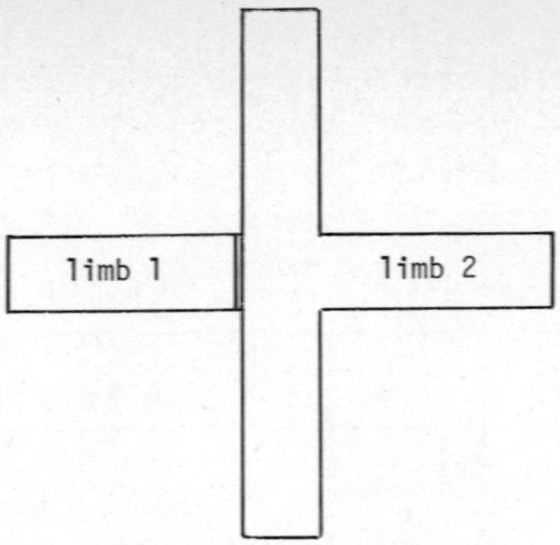

Figure 3.4 Cross-shaped region

from the right in Figure 3.4. At a later date, this "generated element" can be reincorporated into the frontal matrix preferably when some of its variables are now fully-summed and hence can be eliminated. We can, of course, at any instant have several "generated elements" waiting for reassembly and this gives rise to a class of techniques which we call multi-frontal.

Multi-frontal techniques can be extended to solve arbitrary systems efficiently, the main difficulty lying in the organisation of the multi-fronts, that is the transfer of "generated elements" and original elements or equations to and from the frontal matrix.

This discussion parallels that in Duff (1981c), for example, on the use of generalized elements in obtaining fast orderings for symmetric positive definite systems and Duff and Reid (1981a) have combined analysis and factorization routines based on this approach into a package, MA27, in the Harwell Subroutine Library.

Notice, in Figure 3.4, that it is immaterial in terms of the fill-in or the number of multiplications in the factorization, whether we choose pivots from left to right in limb 1, store the "generated element" corresponding to variables on the double line, and then choose pivots from right to left in limb 2 or whether we choose pivots alternately from the ends of limbs 1 and 2. If we use a strict minimum degree criterion (that is we choose, as pivot at each stage, the diagonal entry in the row of the reduced matrix with least non-zeros) then the latter ordering (plus, of course, concurrent elimination along the other limbs) would be obtained. However, for better multi-frontal organisation we prefer orderings of the former kind.

Generalizations of this and the efficient implementation of a multi-frontal scheme are given in detail by Duff and Reid (1981a) but are not important in the context of this paper. The important feature of our multi-frontal code is that the current frontal matrix and any generated elements are held as full matrices and all elimination operations are performed with full matrix code. As in the case of the uni-frontal scheme numerical criteria for actual pivot selection can be employed within the frontal matrix.

In our MA27 code we have included an option for using block pivots of order two so that we can solve symmetric indefinite systems in a stable way. Because our code handles the non positive definite case we compare an experimental version of it with both an unsymmetric solver as well as a positive definite solver in Table 3.2.

Order	147	1176	292	130
Non-zeros	1298	9864	1250	713
Ordering time				
MA27	.016	.119	.053	.056
MA28*	.859	3.400	.534	.170
YSMP	.032	.253	.054	.273
Factorization time				
MA27	.063	.355	.074	.023
MA28	.198	1.207	.137	.044
YSMP	.042	.270	.033	.008
Solution time				
MA27	.010	.052	.015	.005
MA28	.010	.048	.012	.004
YSMP	.007	.033	.008	.003

Table 3.2 The performance of a multi-frontal code. Times are in seconds on an IBM 3033

*Time includes factorization also

MA28 is the Harwell code for general unsymmetric systems (Duff (1977)) and YSMP the Yale code for symmetric positive definite matrices (Eisenstat et al (1977)). We see that our solution times are comparable with MA28 and our factorization times lie midway between that for positive definite systems and that for general matrices. Our ordering times are vastly superior to those for MA28 because we do not need to consider numerical values at this stage and are better than YSMP because of a facility for recognising when one generated element is embedded in another and the way we identify rows which become symbolically identical during the elimination.

4. Vectorization

We illustrated, in Table 1.3, the substantial factor (about 15) by which full matrix code, written in Fortran, speeds up when run on the CRAY-1 because of vectorization of the innermost loops (further gains may be expected if some assembly language programming is done). We now examine the techniques discussed in sections 2 and 3 to see if any of these can reap similar benefits.

We show the results of performing identical runs to those in Table 2.1 but on the CRAY-1 computer in Table 4.1. There are two effects particularly worthy of

Density of active matrix when switch to full code occurs (in percentages)	Order Non-zeros	147 1298	1176 9864	292 1250	130 713
	Time for factorization				
10		.085	.651	.092	.047
20		.062	.501	.081	.041
30		.069	.376	.084	.041
40		.076	.405	.087	.041
50		.081	.393	.089	.041
60		.087	.491	.090	.041
70		.093	.500	.093	.041
80		.098	.499	.093	.041
90		.105	.497	.095	.041
100		.112	.496	.097	.042
no switch		.130	.845	.101	.045

Table 4.1 The effect of switching to full code when running on the CRAY-1. Times in seconds on the CRAY-1.

note. The first is that the benefits of switching to full code are apparent at even lower densities than in the non-vector machine runs. For example, on the 1176 case, the switch over is beneficial only to densities of 80% on the IBM 3033 while with the CRAY we continue gaining in overall time until the density of the reduced matrix is 30%. The other effect is that the vectorization of full code vis-a-vis sparse code is evident in the relative ratios of speeds for switch over at different densities. On the 1176 case, the ratio 3033/CRAY is 2.4 when no full code is used (some of the code has vectorized but not the innermost loop) while at 40% density switchover the ratio is 3.6 and at 20% it is 7.9.

Our frontal code also shows the effect of vectorization. When the innermost loop dominates (in the larger case) the effect is most marked. We show the results from runs on the CRAY-1 in Table 4.2. These times should be compared with those in Table 3.1. Here the ratio in the largest case shown is about 8, still somewhat below the value 15 for full code (Tables 1.1 and 1.3) on the problem as a whole. However, in this example, much time is spent in inputting and outputting to backing store, a type of computation which the CRAY does not do particularly efficiently.

Grid m / n	10 / 10	10 / 40	10 / 60	10 / 100	10 / 300	32 / 32	64 / 64
Decomposition time							
MA32	17	70	106	177	532	370	2886
MA28*	55	300	410	730	2700	4400	-
MA28**	16	70	100	160	530	530	-
Solution time							
MA32	4	14	20	33	99	80	595
MA28	2	7	10	20	60	30	160+

Table 4.2 Performance of frontal code on the CRAY-1. Times in milliseconds on the CRAY-1.

*Pivot ordering and factorization

**Factorization when pivot sequence is known

+Estimated time

The speed up of our multi-frontal code shown in Table 4.3 is a little disappointing. Although the inner-loop does vectorize, it is not dominating the computation to the same extent as in the uni-frontal scheme and so the overall effect on factorization time is not very dramatic, a speed up of 2.8 being attained (see Table 3.2 for a comparison) which is only a slightly greater factor

Order / Non-zeros	147 / 1298	1176 / 9864	292 / 1250	130 / 713
Ordering time				
MA27	.011	.077	.033	.047
MA28*	.475	1.800	.285	.103
YSMP	.019	.134	.031	.175
Factorization time				
MA27	.022	.110	.032	.012
MA28	.085	.580	.059	.020
YSMP	.020	.125	.016	.005
Solution time				
MA27	.002	.019	.006	.002
MA28	.005	.026	.005	.002
YSMP	.003	.016	.004	.001

Table 4.3 The performance of a multi-frontal code on the CRAY-1. Times are in seconds.

than for strictly scalar codes. Similarly, in the times for solution, only in the 147 case, where the inner loop is dominating, is a significant increase in speed (a factor of 5) recorded. These gains will be much more noticeable when larger problems are solved or when our, as yet experimental, MA27 code is tuned further so that the inner-loop computation is more dominant.

We will report on this more fully and on the benefits of assembly coding of inner loops in a forthcoming report (Duff and Reid (1981b)).

References

Cliffe, K.A., Jackson, C.P., Rae, J. and Winters, K.H. (1978). Finite element flow modelling using velocity and pressure variables. Harwell Report, AERE R.9202.

Du Croz, J.J., Nugent, S.M., Reid, J.K. and Taylor, D.B. (1978). Solving large full sets of linear equations in a paged virtual store. Harwell Report, CSS 68. To appear in TOMS.

Duff, I.S. (1977). MA28 - a set of Fortran subroutines for sparse unsymmetric linear equations. Harwell Report, AERE R.8730. HMSO, London.

Duff, I.S. (1981a). MA32 - a package for solving sparse unsymmetric systems using the frontal method. Harwell Report, AERE R.10079. HMSO, London.

Duff, I.S. (1981b). The design and use of a frontal scheme for solving sparse unsymmetric equations. To appear in Proceedings of 3rd Workshop on Numerical Analysis. Cocoyoc, Mexico. Lecture Notes in Mathematics, Springer.

Duff, I.S. (1981c). A sparse future. pp.1-29 in Sparse Matrices and their Uses. I.S. Duff (Ed.). Academic Press.

Duff, I.S. and Reid, J.K. (1979). Performance evaluation of codes for sparse matrix problems. pp.121-135. in Performance Evaluation of Numerical Software. L. Fosdick (Ed.). North Holland.

Duff, I.S. and Reid, J.K. (1981a). MA27 - a set of Fortran subroutines for the solution of sparse symmetric linear equations. To appear as an AERE R report.

Duff, I.S. and Reid, J.K. (1981b). Experience of sparse matrix codes on the CRAY-1. Invited paper at Conference on Vector and Parallel Processors in Computational Science. Chester. August 1981.

Eisenstat, S.C., Gursky, M.C., Schultz, M.H. and Sherman, A.H. (1977). Yale sparse matrix package. I. The Symmetric Codes. Dept. Computer Science, Yale University Report ≠112.

George, A., Liu, J.W.H. and Ng, E. (1980). User Guide for SPARSPAK: Waterloo Sparse Linear Equations Package. Comput. Sci. Report CS-78-30, Waterloo, Canada.

George, A. and Liu, J.W.H. (1981). Computer solution of large sparse positive definite systems. Prentice-Hall.

Hood, P. (1976). Frontal solution program for unsymmetric matrices. Int. J. Numer. Meth. Engng. 10, pp.379-399.

IBM (1976). IBM System/360 and System/370 IBM 1130 and IBM 1800 Subroutine Library - Mathematics. User's Guide. Program Product 5736-XM7. IBM catalogue ≠SH12-5300-1.

Irons, B.M. (1970). A frontal solution program for finite element analysis. Int. J. Numer. Meth. Engng. 2, pp.5-32.

Munksgaard, N. (1980). Solving sparse symmetric sets of linear equations by preconditioned conjugate gradients. TOMS 6, pp.206-219.

SECOND ORDER CORRECTIONS FOR NON-DIFFERENTIABLE OPTIMIZATION

R. Fletcher

1. Introduction

This paper describes methods for solving unconstrained non-differentiable optimization (NDO) problems. A general framework which includes many cases of practical importance is to seek a local minimizer $\underset{\sim}{x}^*$ of the composite function

$$\phi(\underset{\sim}{x}) \stackrel{\Delta}{=} f(\underset{\sim}{x}) + h(\underset{\sim}{c}(\underset{\sim}{x})) \tag{1.1}$$

where $f(\underset{\sim}{x})$ $(\mathbb{R}^n \to \mathbb{R})$ and $\underset{\sim}{c}(\underset{\sim}{x})$ $(\mathbb{R}^n \to \mathbb{R}^m)$ are smooth functions ($\mathbb{C}^1$ at least) and $h(\underset{\sim}{c})$ $(\mathbb{R}^m \to \mathbb{R})$ is convex but nonsmooth $(\mathbb{C}^0)$. A particularly important application of this type is to minimize an exact penalty function arising from a nonlinear programming problem, when $h(\underset{\sim}{c})$ becomes either $||\underset{\sim}{c}||$ or $||\underset{\sim}{c}^+||$, and most commonly the ℓ_1 norm is used. Other applications include finite min-max, ℓ_p approximation for $1 \leq p \leq \infty$ and feasible points of systems of equations and inequalities. These applications are all typified by $h(\underset{\sim}{c})$ being some type of polyhedral convex function

$$h(\underset{\sim}{c}) = \max_{1\leq i\leq p} \underset{\sim}{c}^T\underset{\sim}{h}_i + b_i \tag{1.2}$$

formed from a finite number of supporting hyperplanes. The vectors $\underset{\sim}{h}_i$ and constants b_i are given, but differ depending on the type of application. More details are given by Fletcher (1981a).

An algorithm for this problem is given by Fletcher (1980a, 1981a, 1981b) and is summarized in this section. It is globally convergent under mild conditions through the use of a step restriction or trust region strategy. However there are difficulties for the algorithm if it encounters a 'steep sided curved groove' in the function (1.1), that is to say a nonlinear curve in the domain, across which $\phi(\underset{\sim}{x})$ has a large jump discontinuity (with sign change) of derivative. This can cause slow convergence in general, and also gives rise to the 'Maratos effect' described in section 2. This can obviate the second order rate of convergence of the method which would otherwise occur.

This paper describes a modification to the method based on the possible use of a 'second order correction' or 'projection step'. This is motivated in section 2 where it is shown how use of the correction can improve the asymptotic properties of the method and hence avoid the Maratos effect. In section 3 various practical considerations regarding the use of the correction are discussed, leading to the suggestion of a modified algorithm. It is shown that this algorithm retains the global convergence properties of the unmodified algorithm. In section 4 numerical experience with the method is described showing that the modified algorithm follows curved grooves much more rapidly than the unmodified algorithm, yet handles well behaved problems equally effectively. Finally the possibility of being able to improve the theoretical properties of the method any further is discussed.

The theoretical background required to follow all aspects of this paper is extensive, so frequent reference is made to Fletcher (1981a) for these details. However it is hoped that the main ideas of the paper can be understood with only a minimum amount of this background. First order necessary conditions for $\underset{\sim}{x}^*$ to minimize (1.1) locally are that a vector of Lagrange multipliers $\underset{\sim}{\lambda}^* \in \partial h^*$ exists such that

$$\underset{\sim}{g}^* + A^* \underset{\sim}{\lambda}^* = \underset{\sim}{0} \tag{1.3}$$

where $\underset{\sim}{g} = \underset{\sim}{\nabla} f$, $A = \underset{\sim}{\nabla} \underset{\sim}{c}^T$ (the Jacobian of $\underset{\sim}{c}$) , ∂h denotes the subdifferential of $\underset{\sim}{c}$, and $\underset{\sim}{g}^*$ denotes $\underset{\sim}{g}(\underset{\sim}{x}^*)$ etc. Equation (1.3) expresses the non-existence of descent directions at $\underset{\sim}{x}^*$. For an exact penalty function application the Lagrange multipliers $\underset{\sim}{\lambda}^*$ are closely related to those which exist at the solution to the associated nonlinear programming problem.

The set $\partial h(\underset{\sim}{c})$ is bounded in general if $\underset{\sim}{c}$ is bounded, although for all the common applications there is an a-priori bound; for example for a polyhedral convex function (1.2)

$$\partial h(\underset{\sim}{c}) = \underset{i \in A}{\text{conv}}\ \underset{\sim}{h}_i \tag{1.4}$$

where

$$A = \{i : h(\underset{\sim}{c}) = \underset{\sim}{c}^T \underset{\sim}{h}_i + b_i\} \tag{1.5}$$

is the set of active hyperplanes at $\underset{\sim}{c}$. This implies that the Lagrange multipliers at the minimizer of a composite function are bounded, and this result plays an important part in the global convergence of the methods described here.

The basis of this paper is an algorithm suggested by Fletcher (1980a, 1981a, 1981b) in which a sequence of iterates $\underset{\sim}{x}^{(k)}$, $\underset{\sim}{\lambda}^{(k)}$ are calculated from a given estimate $\underset{\sim}{x}^{(1)}$, $\underset{\sim}{\lambda}^{(1)}$, hopefully converging to $\underset{\sim}{x}^*$, $\underset{\sim}{\lambda}^*$. About $\underset{\sim}{x}^{(k)}$, a linear Taylor series approximation $\underset{\sim}{\ell}^{(k)}(\underset{\sim}{\delta})$ to $\underset{\sim}{c}(\underset{\sim}{x})$ is

$$c(\underset{\sim}{x}^{(k)} + \underset{\sim}{\delta}) \simeq \underset{\sim}{\ell}^{(k)}(\underset{\sim}{\delta}) \triangleq \underset{\sim}{c}^{(k)} + A^{(k)T}\underset{\sim}{\delta} \tag{1.6}$$

where $\underset{\sim}{\delta}$ denotes a potential correction to $\underset{\sim}{x}^{(k)}$. Also a quadratic function $q^{(k)}(\underset{\sim}{\delta})$ which approximates $f(\underset{\sim}{x})$ is

$$f(\underset{\sim}{x}^{(k)} + \underset{\sim}{\delta}) \simeq q^{(k)}(\underset{\sim}{\delta}) \triangleq f^{(k)} + \underset{\sim}{g}^{(k)T}\underset{\sim}{\delta} + \tfrac{1}{2}\underset{\sim}{\delta}^T W^{(k)} \underset{\sim}{\delta} \tag{1.7}$$

where

$$W^{(k)} = \nabla^2 f^{(k)} - \textstyle\sum_i \lambda_i^{(k)} \nabla^2 c_i^{(k)} . \tag{1.8}$$

Expression (1.7) is a second order Taylor series, augmented by terms in (1.8) which account for curvature in the functions $c_i(\underset{\sim}{x})$. By replacing f and $\underset{\sim}{c}$ in (1.1) by (1.6) and (1.7) the composite function

$$\psi^{(k)}(\underset{\sim}{\delta}) = q^{(k)}(\underset{\sim}{\delta}) + h(\underset{\sim}{\ell}^{(k)}(\underset{\sim}{\delta})) \tag{1.9}$$

is defined which can therefore be regarded as an approximation to $\phi(\underset{\sim}{x}^{(k)} + \underset{\sim}{\delta})$. A very simple way of using (1.9), referred to as the

QL method, is to choose $\underset{\sim}{\delta}^{(k)}$ to minimize (1.9) and to set $\underset{\sim}{x}^{(k+1)} = \underset{\sim}{x}^{(k)} + \underset{\sim}{\delta}^{(k)}$ and $\underset{\sim}{\lambda}^{(k+1)}$ as the multipliers associated with the minimizer of (1.9). It is important that the subproblem of minimizing (1.9) is computationally tractable, and this is usually so when $h(\underset{\sim}{c})$ is polyhedral in which case it can be reduced to a quadratic programming (QP) problem. In fact when the ℓ_1 norm is involved then it is more efficient to handle the problem directly (I refer to this as an 'ℓ_1QP problem') based on methods like those of Conn and Sinclair (1975). The definition (1.8) of $W^{(k)}$ is the 'correct' way to take account of second order terms in that it gives rise to a second order rate of convergence in most practical applications under mild conditions (see Fletcher, 1981a, and section 2). The method therefore requires second derivatives to be available: however it is quite possible to take over ideas from nonlinear programming in which $W^{(k)}$ is updated by differences in first derivatives, for example Powell (1977). Special cases related to the QL method have been suggested variously in the literature, for example Pshenichnyi (1978) and Han (1981) in the context of min-max approximation. Also for nonlinear programming the QL method (applied to an exact penalty function) is equivalent to the SOLVER (sequential QP) method of Wilson (1963) and variants thereof.

The QL method alone is not satisfactory as a general purpose method since it is not guaranteed to converge, and is undefined when the subproblem has no solution because $\psi^{(k)}(\underset{\sim}{\delta})$ is unbounded. However, if the idea of a step restriction or trust region is introduced, which has been used successfully in smooth unconstrained optimization (e.g. Moré, 1978; Fletcher, 1980b; Sorensen 1980), then the subproblem which is solved is

$$\begin{aligned} &\text{minimize} \quad \psi^{(k)}(\underset{\sim}{\delta}) \\ &\text{subject to} \quad ||\underset{\sim}{\delta}|| \le \rho^{(k)} \ . \end{aligned} \tag{1.10}$$

The parameter $\rho^{(k)}$ is the radius of the feasible region (or trust region) in (1.10) and is intended to measure the region over which the agreement between $\psi^{(k)}(\underset{\sim}{\delta})$ and $\phi(\underset{\sim}{x}^{(k)} + \underset{\sim}{\delta})$ is adequate in a certain sense. The radius $\rho^{(k)}$ is adjusted adaptively from one iteration to the next by monitoring the ratio of actual to predicted reductions in $\phi(\underset{\sim}{x})$. Fletcher (1981a) suggests a specific algorithm (14.5.6) in which this is done; this is referred to as the basic method in the rest of this paper and is the basis of the modification described here in section 3. The norm in (1.10) is arbitrary but for practical convenience the ℓ_∞ norm is used, which preserves QP or ℓ_1QP subproblems. In fact the trust region constraint is then just the intersection of simple bounds, so it is possible to generalize the methods to minimize $\phi(\underset{\sim}{x})$ in (1.1) subject to upper and lower bounds $\underset{\sim}{\ell} \le \underset{\sim}{x} \le \underset{\sim}{u}$ with no significant complication. The basic method has a nice interpretation that the solution of (1.10) picks out 'locally active discontinuities', and if the trust region is active this set may not include all the active discontinuities at the solution. In terms of an exact ℓ_1 penalty function, this is a set of 'locally active constraints', that is constraints which can be zeroed locally. Note that not all constraints (even equations) may be zeroed in the subproblem, in constrast to SOLVER (sequential QP) and Newton's method. This feature contributes to the strong global convergence properties of the basic method. The basic method is also observed to work well in practice when applied to minimizing well behaved exact ℓ_1 penalty functions in nonlinear programming (Fletcher, 1981b).

2. The second order correction

Difficulties arise in solving an NDO problem when the iterates have to follow a steep sided curved groove as described above. The function $\psi^{(k)}(\underset{\sim}{\delta})$ in subproblem (1.10) involves a linearization of the discontinuity and only a limited amount of progress can be made along this linearization if the objective function is to be reduced. The effect becomes more serious as the steepness of the groove is increased (see section 4). An associated difficulty is the 'Maratos effect' in which for some $\underset{\sim}{x}^{(k)}$ (on or close to a curved groove) arbitrarily close to $\underset{\sim}{x}^*$, a unit step of the QL method fails to reduce $\phi(\underset{\sim}{x})$. Thus the trust region modification for forcing a decrease in ϕ has to be invoked. Superlinear convergence of the QL method depends on taking the unit step at every iteration, so it is no longer possible to guarantee superlinear convergence if every iteration is required to reduce $\phi(\underset{\sim}{x})$.

An approach to the solution of these difficulties is motivated by ideas used in feasible direction methods of nonlinear programming. The way in which these methods follow the curved boundary of the feasible region is typified by Figure 1. At a feasible point $\underset{\sim}{x}^{(k)}$, a basic step $\underset{\sim}{\delta}^{(k)}$ is calculated in the tangent plane on the basis of some minimization method, giving a point $\hat{\underset{\sim}{x}} = \underset{\sim}{x}^{(k)} + \underset{\sim}{\delta}^{(k)}$. To restore feasibility a sequence of projection steps is then carried out, giving a new feasible point $\underset{\sim}{x}^{(k+1)}$ for the next major iteration. The first step of the projection iteration is to a point $\tilde{\underset{\sim}{x}} = \hat{\underset{\sim}{x}} + \hat{\underset{\sim}{\delta}}$ and the correction $\hat{\underset{\sim}{\delta}}$ is referred to as the <u>second order correction</u>. This is because asymptotically the basic step cancels out the error $\underset{\sim}{x}^{(k)} - \underset{\sim}{x}^*$ to first order, so that $\hat{\underset{\sim}{\delta}}$ is second order in this error. The second order correction may be calculated by

$$\hat{\underset{\sim}{\delta}} = - S\hat{\underset{\sim}{c}} \tag{2.1}$$

Figure 1

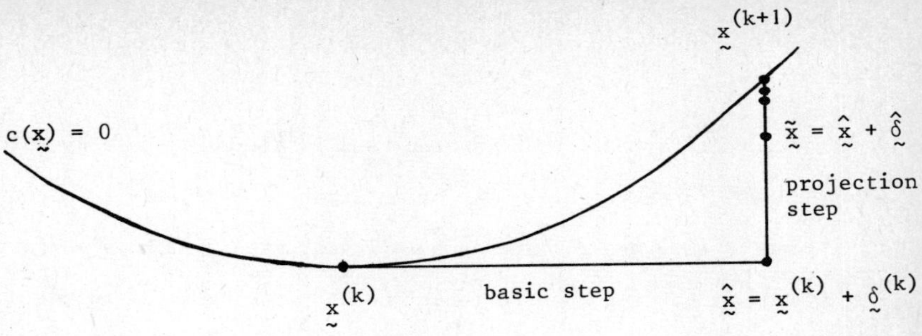

A feasible direction method

where $S^T : S^TA^{(k)} = I$ is any convenient left inverse for the Jacobian matrix at $\underset{\sim}{x}^{(k)}$. Thus only the constraint residuals $\hat{\underset{\sim}{c}}$ at $\hat{\underset{\sim}{x}}$ need be recalculated. Various ways of choosing a suitable S matrix have been suggested: I like one essentially due to Sargent (1974), which also generalizes nicely to an NDO context. This method is motivated by trying to find a feasible correction to a subproblem obtained by replacing $f(\underset{\sim}{x})$ by $q^{(k)}(\underset{\sim}{\delta})$ in the nonlinear programming problem, and incorporating a trust region restriction, that is

$$\begin{aligned} &\underset{\underset{\sim}{\delta}}{\text{minimize}} \quad q^{(k)}(\underset{\sim}{\delta}) \\ &\text{subject to} \quad \underset{\sim}{c}(\underset{\sim}{x}^{(k)} + \underset{\sim}{\delta}) = \underset{\sim}{0} \\ &\qquad\qquad ||\underset{\sim}{\delta}|| \le \rho^{(k)} . \end{aligned} \tag{2.2}$$

This problem is nonlinear but may be solved by repeated linearization of the constraint functions, evaluating $\underset{\sim}{c}$ anew at each point, but using derivatives evaluated at $\underset{\sim}{x}^{(k)}$. The first step of this process is the basic step referred to above, and the second step is a second order correction.

Second order corrections have been used in an NDO context by Coleman and Conn (1980) in the special case of an exact ℓ_1 penalty function. In this paper a more general approach is investigated which is appropriate to the minimization of composite functions and fits easily into the framework of the QL method. By analogy with (2.2), $f(\underset{\sim}{x})$ in (1.1) is replaced by $q^{(k)}(\underset{\sim}{\delta})$ and a step restriction is added, but the $\underset{\sim}{c}(\underset{\sim}{x})$ term is left unchanged, giving the subproblem

$$\begin{aligned} &\underset{\underset{\sim}{\delta}}{\text{minimize}} \quad q^{(k)}(\underset{\sim}{\delta}) + h(\underset{\sim}{c}(\underset{\sim}{x}^{(k)} + \underset{\sim}{\delta})) \\ &\text{subject to} \quad ||\underset{\sim}{\delta}|| \le \rho^{(k)} . \end{aligned} \tag{2.3}$$

Because $\underset{\sim}{c}(\underset{\sim}{x})$ is nonlinear, this subproblem cannot generally be solved finitely, but linearization of $\underset{\sim}{c}$ about $\underset{\sim}{x}^{(k)}$ as above gives the <u>basic step</u>

$$\begin{aligned} &\text{minimize} \quad \psi^{(k)}(\underset{\sim}{\delta}) \triangleq q^{(k)}(\underset{\sim}{\delta}) + h(\underset{\sim}{c}^{(k)} + A^{(k)T}\underset{\sim}{\delta}) \\ &\text{subject to} \quad ||\underset{\sim}{\delta}|| \le \rho^{(k)} \end{aligned} \tag{2.4}$$

which is solved to give $\underset{\sim}{\delta}^{(k)}$. Then

$$\hat{\underset{\sim}{x}} = \underset{\sim}{x}^{(k)} + \underset{\sim}{\delta}^{(k)} \tag{2.5}$$

is calculated and another linearization of $\underset{\sim}{c}$ about $\hat{\underset{\sim}{x}}$ as above gives the <u>projection step</u>

$$\begin{aligned} &\text{minimize} \quad \hat{\psi}(\underset{\sim}{\delta}) \triangleq q^{(k)}(\underset{\sim}{\delta} + \underset{\sim}{\delta}^{(k)}) + h(\hat{\underset{\sim}{c}} + A^{(k)T}\underset{\sim}{\delta}) \\ &\text{subject to} \quad ||\underset{\sim}{\delta} + \underset{\sim}{\delta}^{(k)}|| \le \rho^{(k)} \end{aligned} \tag{2.6}$$

which is solved for the second order correction $\hat{\underset{\sim}{\delta}}$, and

$$\tilde{\underset{\sim}{x}} = \hat{\underset{\sim}{x}} + \hat{\underset{\sim}{\delta}} . \tag{2.7}$$

An algorithm has been devised based on solving these subproblems and is described in more detail in section 3. Either $\hat{\underset{\sim}{x}}$ or $\tilde{\underset{\sim}{x}}$ (or $\underset{\sim}{x}^{(k)}$) is taken as $\underset{\sim}{x}^{(k+1)}$ and the multipliers from the corresponding subproblems determine $\underset{\sim}{\lambda}^{(k+1)}$. It is important to observe that both (2.4) and (2.6) are both ℓ_1QP subproblems and hence can share the same computer software. Since the main change from (2.4) to (2.6) is to replace $\underset{\sim}{c}^{(k)}$ by $\hat{\underset{\sim}{c}}$, solution of (2.6) is in the nature of a parametric programming problem, and if the software permits this to be done it will usually be possible to compute the solution of (2.6) directly from the factorizations used in solving (2.4), and this is analogous to the use of the simple formula (2.1) in a nonlinear programming context. Because of this, and because no derivative information at $\hat{\underset{\sim}{x}}$ is required, the projection step (2.6) is both easy to incorporate and cheap to compute.

It is possible to prove (under mild conditions and assuming that the step restriction is inactive) that use of the second order correction avoids the Maratos effect (Theorem 2.2 below). A result of this type is not particularly novel although it is interesting to see here that it can be proved in the very general setting of composite functions. In analysing the asymptotic behaviour it is assumed that $(\underset{\sim}{x}^{(k)}, \underset{\sim}{\lambda}^{(k)}) \to (\underset{\sim}{x}^*, \underset{\sim}{\lambda}^*)$ and that certain mild conditions (mainly second order sufficient conditions, strict complementarity and a linear independence assumption) hold at $\underset{\sim}{x}^*, \underset{\sim}{\lambda}^*$. The errors in $\underset{\sim}{x}^{(k)}$ and $\underset{\sim}{\lambda}^{(k)}$ are defined by

$$\varepsilon_x = ||\underset{\sim}{x}^{(k)} - \underset{\sim}{x}^*||$$

$$\varepsilon_\lambda = ||\underset{\sim}{\lambda}^{(k)} - \underset{\sim}{\lambda}^*||$$

and it is convenient to denote

$$\varepsilon = \max(\varepsilon_x, \varepsilon_\lambda) .$$

By demonstrating that (2.4) and (2.6) are equivalent to sequential QP techniques close to $\underset{\sim}{x}^*, \underset{\sim}{\lambda}^*$ it is first of all shown that

$$\begin{aligned} ||\underset{\sim}{\hat{x}} - \underset{\sim}{x}^*|| &= 0(\varepsilon.\varepsilon_x) \\ ||\underset{\sim}{\hat{\lambda}} - \underset{\sim}{\lambda}^*|| &= 0(\varepsilon_x^2) \end{aligned} \tag{2.8}$$

for the result of the basic step (2.4), and that

$$\begin{aligned} ||\underset{\sim}{\tilde{x}} - \underset{\sim}{x}^*|| &= 0(\varepsilon.\varepsilon_x) \\ ||\underset{\sim}{\tilde{\lambda}} - \underset{\sim}{\lambda}^*|| &= 0(\varepsilon_x^2) \end{aligned} \tag{2.9}$$

for the projection step, as well as obtaining an order of magnitude estimate for $\underset{\sim}{\hat{\delta}}$ and other quantities (Theorem 2.1). A comparison of (2.8) and (2.9) indicates that both the basic and basic+projection steps give rise to a second order rate of convergence. However, the second order correction does not give the solution to a higher order of accuracy than can be achieved with the basic step. It is therefore instructive to consider exactly what advantage the projection step does confer. In fact if first derivatives are evaluated at $\underset{\sim}{\hat{x}}$ and used in (2.6) then it can be shown that the combination of basic + projection step has a third order rate of convergence (e.g. Wolfe, 1978), but this would be at the cost of a substantial increase in the computation required.

The advantage of using the projection step becomes apparent when considering the extent to which the objective function can be reduced. Lemma 2.2 shows that the maximum reduction that can be achieved is at least a second order quantity, that is there exists a constant $a > 0$ such that

$$\phi^{(k)} - \phi^* \geq a\varepsilon_x^2 \ . \tag{2.10}$$

Furthermore this maximum reduction is approximated to second order by the predicted reduction $\Delta\psi^{(k)}(\triangleq \psi^{(k)}(\underset{\sim}{0}) - \psi^{(k)}(\underset{\sim}{\delta}^{(k)}))$ on the basic step, that is

$$\phi^{(k)} - \phi^* = \Delta\psi^{(k)} + O(\varepsilon.\varepsilon_x^2). \qquad (2.11)$$

The same is true for the actual reduction achieved on the basic + projection step

$$\phi^{(k)} - \tilde{\phi} = \Delta\psi^{(k)} + O(\varepsilon.\varepsilon_x^2) , \qquad (2.12)$$

and these results are given in Lemmas 2.5 and 2.6. It is important to emphasize that it is the predicted reduction on the basic step alone which agrees to second order with the actual reduction on the basic + projection step. Thus it follows that

$$\frac{\phi^{(k)} - \tilde{\phi}}{\Delta\psi^{(k)}} = 1 + O(\varepsilon) \qquad (2.13)$$

and hence that $\phi^{(k)} > \tilde{\phi}$ for k sufficiently large. Asymptotically therefore, the unrestricted basic + projection step always reduces ϕ and the Maratos effect cannot occur (Theorem 2.2). On the other hand the actual reduction on the basic step $\phi^{(k)} - \hat{\phi}$ can give errors in the second order terms. Then if the first order terms in $\phi^{(k)} - \phi^*$ are negligible ($\underset{\sim}{x}^{(k)}$ close to a discontinuity) it is possible that

$$\frac{\phi^{(k)} - \hat{\phi}}{\Delta\psi^{(k)}} < 0 \qquad (2.14)$$

which corresponds to the Maratos effect occurring. It is emphasized that these results relate only to the unrestricted algorithm, or to the algorithm of section 3 only when the step restriction is inactive for all k sufficiently large. The extent to which stronger results can be proved is discussed at the end of section 4.

The rest of this section gives details of the proofs of the above results. In doing this, frequent reference is made to material in Fletcher (1981a) and three part references, e.g. Theorem 14.3.1 or equation (12.3.15), are to this book. The starting point is the second order rate of convergence proof for the basic step of the QL method in Fletcher (1981a), pp. 203–204. This result applies in two different types of composite function problem; one in which $h(\underset{\sim}{c})$ is polyhedral, and the other in which $h(\underset{\sim}{c})$ is $||\underset{\sim}{c}||$ (or $||\underset{\sim}{c}^{+}||$) and $\underset{\sim}{c} = \underset{\sim}{0}$ (or $\underset{\sim}{c}^{+} = \underset{\sim}{0}$). The first case is typical of a max function application and the second of an exact penalty function application. In both cases the method of proof is to show that the NDO problem is equivalent to a certain nonlinear programming (NLP) problem, either (14.1.6) or (14.3.4). Then Theorem 12.3.1 (and Question 12.18) shows that the SOLVER method applied to the NLP problem converges at a second order rate. Finally the SOLVER method applied to the NLP problem is shown to be equivalent to the QL method applied to the NDO problem. Thus the rate of convergence of the QL method is established. The assumptions of Theorem 12.3.1 (and Question 12.18) (second order sufficiency, strict complementarily, linear independence) will be referred to for brevity as <u>standard assumptions</u> and are sufficient for the other stages of the proof.

A similar procedure is followed here for analysing the QL method + projection. It is first shown that this is equivalent to the SOLVER method + projection applied to the NLP problem. In the exact penalty function application the SOLVER + projection method solves

$$\begin{aligned} &\text{minimize} \quad q^{(k)}(\underset{\sim}{\delta}^{(k)} + \underset{\sim}{\delta}) \\ &\text{subject to} \quad \hat{\underset{\sim}{c}} + A^{(k)T}\underset{\sim}{\delta} = \underset{\sim}{0} \end{aligned} \qquad (2.15)$$

on the projection step (for an equality constraint problem). The objective function (2.6) (with $h(\underset{\sim}{c}) = ||\underset{\sim}{c}||$) which is solved in the QL method is an exact penalty function for (2.15) so it is clear that they are equivalent. In the polyhedral function type of application the SOLVER method + projection solves

$$\begin{aligned} &\underset{\delta, v}{\text{minimize}} \quad v + \tfrac{1}{2}(\delta^{(k)} + \delta)^T W^{(k)}(\delta^{(k)} + \delta) + g^{(k)T}\delta^{(k)} + f^{(k)} - \hat{f} \\ &\text{subject to} \quad v - (g^{(k)} + A^{(k)}h_i)^T\delta \geq \hat{f} + \hat{c}^T h_i + b_i \, . \end{aligned} \tag{2.16}$$

on the projection step (see (14.4.10)), where the constant term $g^{(k)T}\delta^{(k)} + f^{(k)} - \hat{f}$ has been added to the objective function without changing the solution. Writing the objective function as w, (2.16) can be written

$$\begin{aligned} &\underset{\delta, w}{\text{minimize}} \quad w \\ &\text{subject to} \quad w \geq q^{(k)}(\delta^{(k)} + \delta) + h_i^T(\hat{c} + A^{(k)T}\delta) + b_i \end{aligned} \tag{2.17}$$

which is equivalent to (2.6) when the step restriction is inactive. Thus the rate of convergence for the QL method + projection in the two cases of interest here, can be determined by analysing the SOLVER method + projection applied to a nonlinear programming problem. This is done below in regard to the equality problem (14.3.1) : the results apply equally to the inequality problem (14.3.3) because of the assumption of strict complementarity.

The basic SOLVER step is defined by

$$\begin{bmatrix} W^{(k)} & A^{(k)} \\ A^{(k)T} & 0 \end{bmatrix} \begin{pmatrix} \delta^{(k)} \\ \hat{\lambda} \end{pmatrix} = - \begin{pmatrix} g^{(k)} \\ c^{(k)} \end{pmatrix} \tag{2.18}$$

and as in Theorem 12.3.1 it follows that

$$\begin{pmatrix} \hat{x} - x^* \\ \hat{\lambda} - \lambda^* \end{pmatrix} = \begin{pmatrix} O(\varepsilon.\varepsilon_x) \\ O(\varepsilon_x^2) \end{pmatrix} , \tag{2.19}$$

which proves (2.8). It also follows that

$$||\delta^{(k)}|| = \varepsilon_x(1 + O(\varepsilon)) \tag{2.20}$$

$$||\hat{\lambda} - \lambda^{(k)}|| = \varepsilon_\lambda + O(\varepsilon_x^2) \, . \tag{2.21}$$

Then by Taylor series

$$\hat{c}_i = c_i^{(k)} + a_i^{(k)T}\delta^{(k)} + O(||\delta^{(k)}||^2) = O(\varepsilon_x^2) \tag{2.22}$$

from (2.18). The equations which are solved in the projection step are

$$
\begin{bmatrix} W^{(k)} & A^{(k)} \\ A^{(k)T} & 0 \end{bmatrix} \begin{pmatrix} \hat{\underset{\sim}{\delta}} \\ \underset{\sim}{\tilde{\lambda}} \end{pmatrix} = - \begin{pmatrix} g^{(k)} + W^{(k)} \underset{\sim}{\delta}^{(k)} \\ \hat{\underset{\sim}{c}} \end{pmatrix}
$$
(2.23)

or using (2.18)

$$
\begin{bmatrix} W^{(k)} & A^{(k)} \\ A^{(k)T} & 0 \end{bmatrix} \begin{pmatrix} \hat{\underset{\sim}{\delta}} \\ \underset{\sim}{\tilde{\lambda}} - \hat{\underset{\sim}{\lambda}} \end{pmatrix} = \begin{pmatrix} 0 \\ -\hat{\underset{\sim}{c}} \end{pmatrix} .
$$
(2.24)

The inverse of the Lagrangian matrix is bounded (e.g. Question 12.4) and so from (2.22) and (2.18)

$$
\hat{\underset{\sim}{\delta}} = O(\varepsilon_x^2) \quad \text{and} \quad \underset{\sim}{\tilde{\lambda}} - \hat{\underset{\sim}{\lambda}} = O(\varepsilon_x^2) ,
$$
(2.25)

from which (2.9) follows. These results hold directly for the exact penalty function application : for the max function type these results are true for the multipliers $\underset{\sim}{\mu}$ in (14.1.7). Since however $\underset{\sim}{\lambda} = H\underset{\sim}{\mu}$ they also hold for the multipliers $\underset{\sim}{\lambda}$. Thus the following has been proved.

Theorem 2.1   For the unrestricted QL method + projection applied to both types of problem under consideration, given standard assumptions, then the method converges at a second order rate, and (2.8), (2.9), (2.19), (2.20), (2.21) and (2.25) all hold.   ☐

The error estimates for function reductions can now be derived in a number of lemmas leading to a proof of (2.13) (Theorem 2.2).

Lemma 2.1   The set G^* in (14.2.20) is closed.

Proof   Let $\{\underset{\sim}{s}_k\} \to \underset{\sim}{s}_\infty$ be a convergent sequence in G^* , and let $\max_{\underset{\sim}{g} \in \partial\phi^*} \underset{\sim}{g}^T \underset{\sim}{s}_\infty = \mu \neq 0$. Since $\partial\phi^*$ is compact there exist vectors, $\underset{\sim}{g}_k$ say, in $\partial\phi^*$ which achieve $\max_{\underset{\sim}{g} \in \partial\phi^*} \underset{\sim}{g}^T \underset{\sim}{s}_k$ and the sequence $\{\underset{\sim}{g}_k\}$ has an accumulation point $\underset{\sim}{g}_\infty$.   By (14.2.20) $\underset{\sim}{g}_k^T \underset{\sim}{s}_k = 0$ and so $\underset{\sim}{g}_\infty^T \underset{\sim}{s}_\infty = 0$. Thus $\mu > 0$. Now $\underset{\sim}{g}_\infty^T \underset{\sim}{s}_k \leq 0$ by definition of $\underset{\sim}{s}_k$ which implies $\underset{\sim}{g}_\infty^T \underset{\sim}{s}_\infty \leq 0$ and $\mu \leq 0$ which is a contradiction. Thus $\mu = 0$, $\underset{\sim}{s}_\infty \in G^*$ and so G^* is closed.

                                                                          ☐

Lemma 2.2 For all $\underset{\sim}{x}^{(k)}$ in some neighbourhood of $\underset{\sim}{x}^*$, there exists a constant $a > 0$ such that $\phi^{(k)} - \phi^* \geq a\varepsilon_x^2$.

Proof Assuming second order conditions (Theorem 14.2.3) then by Lemma 2.1 $\exists$ a constant $a > 0$ such that $\underset{\sim}{s}^T W^* \underset{\sim}{s} \geq 4a \quad \forall \quad \underset{\sim}{s} \in G^*$.

Define $\tilde{\phi}(\underset{\sim}{x}) = \phi(\underset{\sim}{x}) - a||\underset{\sim}{x} - \underset{\sim}{x}^*||_2^2$. It follows as in the proof of Theorem 14.2.3 that $\tilde{\phi}(\underset{\sim}{x})$ has an isolated local minimizer at $\underset{\sim}{x}^*$. Thus the Lemma is proved and the inequality (2.10) is justified. □

Lemma 2.3 For the types of problem under consideration, given standard assumptions, it follows for k sufficiently large that

$$\partial h(\underset{\sim}{c}^*) = \partial h(\underset{\sim}{\ell}^{(k)}(\underset{\sim}{\delta}^{(k)})) = \partial h(\hat{\underset{\sim}{\ell}}(\hat{\underset{\sim}{\delta}})) .$$

Proof The equivalence of the QL method + projection to SOLVER + projection is used. For a polyhedral type problem standard assumptions ensure that the active set A^* in (14.1.10) is the active set when solving (14.4.10) and (2.16), asymptotically. Thus by (14.1.9) the Lemma follows. For an exact inequality penalty function problem (14.3.3), $\underset{\sim}{c}^{*+} = \underset{\sim}{0}$ and (14.3.8) gives

$$\partial h(\underset{\sim}{c}^*) = \{\underset{\sim}{\lambda} : \underset{\sim}{\lambda} \geq \underset{\sim}{0} , c_i^* < 0 \Rightarrow \lambda_i = 0 , ||\underset{\sim}{\lambda}||_D \leq 1\} .$$

Asymptotically for the equivalent QP subproblem it follows from standard assumptions that $(\underset{\sim}{\ell}^{(k)}(\underset{\sim}{\delta}^{(k)}))^+ = (\hat{\underset{\sim}{\ell}}(\hat{\underset{\sim}{\delta}}))^+ = \underset{\sim}{0}$, and $c_i^* < 0 \Leftrightarrow$ $\ell_i^{(k)}(\underset{\sim}{\delta}^{(k)}) < 0 \Leftrightarrow \hat{\ell}_i(\hat{\underset{\sim}{\delta}}) < 0$. Thus the Lemma holds. A similar proof holds for an equality constraint problem. □

Lemma 2.4 For any $\underset{\sim}{\lambda} \in \partial h(\underset{\sim}{c}^*)$

$$h(\underset{\sim}{c}^*) - \underset{\sim}{\lambda}^T\underset{\sim}{c}^* = h(\underset{\sim}{\ell}^{(k)}(\underset{\sim}{\delta}^{(k)})) - \underset{\sim}{\lambda}^T\underset{\sim}{\ell}^{(k)}(\underset{\sim}{\delta}^{(k)}) = h(\hat{\underset{\sim}{\ell}}(\hat{\underset{\sim}{\delta}})) - \underset{\sim}{\lambda}^T\hat{\underset{\sim}{\ell}}(\hat{\underset{\sim}{\delta}})) .$$

Proof For a polyhedral type problem, because A^* is asymptotically the active set for (14.4.10) and (2.16), it follows that if $\underset{\sim}{\lambda} = \sum_{i\in A^*}\underset{\sim}{h}_i\mu_i$ then from (14.1.10)

$$h(\underset{\sim}{c}^*) = \underset{\sim}{c}^{*T}\underset{\sim}{\lambda} + \sum b_i\mu_i$$

and

$$h(\underset{\sim}{\ell}^{(k)}(\underset{\sim}{\delta}^{(k)})) = \underset{\sim}{\ell}^{(k)}(\underset{\sim}{\delta}^{(k)})^T\underset{\sim}{\lambda} + \sum b_i\mu_i$$

and

$$h(\hat{\underset{\sim}{\ell}}(\hat{\underset{\sim}{\delta}})) = \hat{\underset{\sim}{\ell}}(\hat{\underset{\sim}{\delta}})^T\underset{\sim}{\lambda} + \sum b_i\mu_i \;.$$

Thus the Lemma holds. For any exact penalty function type problem the result follows from the condition $\underset{\sim}{\lambda}^T\underset{\sim}{c} = h(\underset{\sim}{c})$ in (14.3.7) and (14.3.8), and Lemma 2.3. □

<u>Lemma 2.5</u> $\phi^{(k)} - \phi^* = \Delta\psi^{(k)} + O(\varepsilon.\varepsilon_x^2)$

Defining $\underset{\sim}{e}^{(k)} = \underset{\sim}{x}^{(k)} - \underset{\sim}{x}^*$, $\hat{\underset{\sim}{e}} = \hat{\underset{\sim}{x}} - \underset{\sim}{x}^*$ and using Taylor series about $\underset{\sim}{x}^*$ and Lemmas 2.3 and 2.4, and then (2.21) and $\underset{\sim}{e}^{(k)} + \underset{\sim}{\delta}^{(k)} = \hat{\underset{\sim}{e}}$, it follows that

$$\begin{aligned}
\phi^* &= f^* + h(\underset{\sim}{c}^*) \\
&= f^{(k)} - \underset{\sim}{g}^{(k)T}\underset{\sim}{e}^{(k)} + \tfrac{1}{2}\underset{\sim}{e}^{(k)T}G^{(k)}\underset{\sim}{e}^{(k)} + \sum \hat{\lambda}_i(c_i^{(k)} - \underset{\sim}{a}_i^{(k)}\underset{\sim}{e}^{(k)} + \tfrac{1}{2}\underset{\sim}{e}^{(k)T}G_i^{(k)}\underset{\sim}{e}^{(k)}) \\
&\qquad + h(\underset{\sim}{\ell}^{(k)}(\underset{\sim}{\delta}^{(k)})) - \hat{\underset{\sim}{\lambda}}^T(\underset{\sim}{c}^{(k)} + A^{(k)T}\underset{\sim}{\delta}^{(k)}) + O(\varepsilon_x^3) \\
&= f^{(k)} - \underset{\sim}{g}^{(k)T}\underset{\sim}{e}^{(k)} + \tfrac{1}{2}\underset{\sim}{e}^{(k)T}W^{(k)}\underset{\sim}{e}^{(k)} - \hat{\underset{\sim}{\lambda}}^TA^{(k)}\hat{\underset{\sim}{e}} + h(\underset{\sim}{\ell}^{(k)}(\underset{\sim}{\delta}^{(k)})) + O(\varepsilon.\varepsilon_x^2) \\
&= f^{(k)} + \underset{\sim}{g}^{(k)T}\underset{\sim}{\delta}^{(k)} + \tfrac{1}{2}\underset{\sim}{\delta}^{(k)T}W^{(k)}\underset{\sim}{\delta}^{(k)} + h(\underset{\sim}{\ell}^{(k)}(\underset{\sim}{\delta}^{(k)})) + O(\varepsilon.\varepsilon_x^2)
\end{aligned}$$

by $\underset{\sim}{g}^{(k)} + W^{(k)}\underset{\sim}{\delta}^{(k)} + A^{(k)}\hat{\underset{\sim}{\lambda}} = \underset{\sim}{0}$ (first order conditions for unrestricted (2.4)) and (2.19). [Note $G = \nabla^2 f$ and $G_i = \nabla^2 c_i$.] But

$$\begin{aligned}
\Delta\psi^{(k)} &= \phi^{(k)} - \psi^{(k)}(\underset{\sim}{\delta}^{(k)}) \\
&= \phi^{(k)} - f^{(k)} - \underset{\sim}{g}^{(k)T}\underset{\sim}{\delta}^{(k)} - \tfrac{1}{2}\underset{\sim}{\delta}^{(k)T}W^{(k)}\underset{\sim}{\delta}^{(k)} - h(\underset{\sim}{\ell}^{(k)}(\underset{\sim}{\delta}^{(k)}))
\end{aligned}$$

Combining the two results proves the Lemma. □

<u>Lemma 2.6</u> $\phi^{(k)} - \tilde{\phi} = \Delta\psi^{(k)} + O(\varepsilon.\varepsilon_x^2)$

<u>Proof</u> Using a Taylor series for $\tilde{\underset{\sim}{c}}$ about $\hat{\underset{\sim}{x}}$, then a Taylor series for $\hat{f}$ about $\underset{\sim}{x}^{(k)}$ using (2.20) and (2.25) and the bound on ∂h, and then Lemma 2.4, it follows that

$$\begin{aligned}
\tilde{\phi} &= \tilde{f} + h(\tilde{\underset{\sim}{c}}) \\
&= \tilde{f} + h(\hat{\underset{\sim}{\ell}}(\hat{\underset{\sim}{\delta}})) + O(\|\hat{\underset{\sim}{\delta}}\|^2)) \\
&= f^{(k)} + \underset{\sim}{g}^{(k)T}(\underset{\sim}{\delta}^{(k)} + \hat{\underset{\sim}{\delta}}) + \tfrac{1}{2}\underset{\sim}{\delta}^{(k)T}G^{(k)}\underset{\sim}{\delta}^{(k)} + h(\hat{\underset{\sim}{\ell}}(\hat{\underset{\sim}{\delta}})) + O(\varepsilon_x^3)
\end{aligned}$$

$$= f^{(k)} + \underset{\sim}{g}^{(k)T}(\underset{\sim}{\delta}^{(k)} + \hat{\underset{\sim}{\delta}}) + \tfrac{1}{2}\underset{\sim}{\delta}^{(k)T}G^{(k)}\underset{\sim}{\delta}^{(k)} + \sum \hat{\lambda}_i(\hat{c}_i + \hat{\underset{\sim}{a}}_i^T\hat{\underset{\sim}{\delta}})$$

$$+ h(\underset{\sim}{\ell}^{(k)}(\underset{\sim}{\delta}^{(k)})) - \hat{\underset{\sim}{\lambda}}^T(\underset{\sim}{c}^{(k)} + A^{(k)T}\underset{\sim}{\delta}^{(k)}) + O(\varepsilon_x^3)$$

$$= f^{(k)} + \underset{\sim}{g}^{(k)T}\underset{\sim}{\delta}^{(k)} + \tfrac{1}{2}\underset{\sim}{\delta}^{(k)T}W^{(k)}\underset{\sim}{\delta}^{(k)} + h(\underset{\sim}{\ell}^{(k)}(\underset{\sim}{\delta}^{(k)})) + O(\varepsilon.\varepsilon_x^2)$$

after using $\underset{\sim}{g}^{(k)} + W^{(k)}(\underset{\sim}{\delta}^{(k)} + \hat{\underset{\sim}{\delta}}) + A^{(k)}\tilde{\underset{\sim}{\lambda}} = \underset{\sim}{0}$ (first order conditions for unrestricted (2.6)) and a second order Taylor series for $\hat{c}_i$ about $\underset{\sim}{x}^{(k)}$, and coalescing various terms of order $\varepsilon.\varepsilon_x^2$ or lower. As in Lemma 2.5 the result follows. □

Finally it is observed that a similar result for $\phi^{(k)} - \hat{\phi}$ does not hold because $h(\hat{\underset{\sim}{c}}) - h(\underset{\sim}{\ell}^{(k)}(\underset{\sim}{\delta}^{(k)}))$ usually contains terms $\sim \varepsilon_x^2$ which cannot be eliminated.

The main result of this section follows easily from Lemmas 2.2, 2.5 and 2.6 and is the following.

Theorem 2.2 For the unrestricted QL method + projection applied to both types of problem under consideration, given standard assumptions, then $(\phi^{(k)} - \tilde{\hat{\phi}})/\Delta\psi^{(k)} = 1 + O(\varepsilon)$ which is (2.13) and implies that the Maratos effect cannot occur. □

3. An algorithm

In Fletcher (1981a), a basic method (algorithm (14.5.6)) is given which incorporates the basic step of the QL method, and it is shown how the trust region radius $\rho^{(k)}$ can be adjusted adaptively to give a global convergence proof under mild conditions. A practical algorithm based on this has performed well in applications of an exact penalty function (Fletcher, 1981b). The aim of this section is to show how the projection step (2.6) which calculates the second order correction can be incorporated into the algorithm to cater more effectively for steep sided curved grooves and the Maratos effect, whilst retaining the global convergence proof and other good features of the basic method. As in the basic method, decisions for changing $\rho^{(k)}$ are based on measuring agreement between the true and approximate objective functions at any point $\underset{\sim}{x}$ on iteration k by the ratio

$$r(\underset{\sim}{x}) = \frac{\phi^{(k)} - \phi(\underset{\sim}{x})}{\Delta\psi^{(k)}} \qquad (3.1)$$

and $\hat{r} = r(\hat{\underset{\sim}{x}})$ etc. is used. As usual a value close to 1 is regarded as good agreement, a value close to zero as poor agreement, and a negative value corresponds to a worse point at which ϕ is increased. Asymptotically (2.13) shows that the unrestricted QL method + projection gives arbitrarily good agreement.

Examining the convergence proof for the basic method in a more general context, it can be seen that if $\hat{r} < 0.25$ and $r(\underset{\sim}{x}) < 0.25$ at any other point tried on iteration k, then it is allowable to take $\underset{\sim}{x}^{(k+1)}$ as either $\underset{\sim}{x}^{(k)}$ or any better point (lower ϕ) but $\rho^{(k+1)}$ must be set to some fixed fraction of $||\underset{\sim}{\delta}^{(k)}||$. Also when $r(\underset{\sim}{x}) \geq 0.25$ occurs at $\hat{\underset{\sim}{x}}$ or any other point, then $\underset{\sim}{x}^{(k+1)}$ must be set to that point. As long as these features are retained in a more general method, then the proof of global convergence carries over. The main strategy therefore is to increase ρ if good agreement occurs and $||\underset{\sim}{\delta}|| \leq \rho^{(k)}$ is active, and to decrease ρ if poor agreement occurs (both as in the basic method), but with some provisions for trying the projection step. As an indication of whether the projection step is likely to be good, it is possible to use the solution of (2.6) to estimate the additional reduction in ϕ to be achieved. Defining $\Delta\hat{\psi} = \hat{\psi}(\underset{\sim}{0}) - \hat{\psi}(\hat{\underset{\sim}{\delta}})$ then an estimate of $\tilde{r}$ is

$$\tilde{r}_e = \frac{\phi^{(k)} - \hat{\phi} + \Delta\hat{\psi}}{\Delta\psi^{(k)}} = \hat{r} + \Delta\hat{\psi}/\Delta\psi^{(k)} . \qquad (3.2)$$

If poor agreement occurs on the basic step but $\tilde{r}_e$ indicates that good agreement can be obtained by taking the projected step, then this is done.

Some thought was given as to whether to use the projected step if $\hat{r} \in [0.25, 0.75]$ indicates average agreement. Practical considerations indicate that to take the projection step might force the iterates into a groove and slow down the overall rate of convergence: since the basic step has obtained average agreement, use of the projection step is unnecessary. On the other hand if $\underset{\sim}{x}^{(k)}$ is actually in a groove then taking the projection step might increase $r(\tilde{\underset{\sim}{x}})$ to above 0.75 and enable larger steps to be taken when following the groove. This choice was resolved by accepting the basic step in these circumstances but increasing ρ for the next iteration if $\tilde{r}_e$ predicts good agreement of the projection step. Another feature which required some thought is that it is very convenient for the user if any evaluation of derivatives always directly follows an evaluation of the functions at the same point (otherwise common sub-expressions cannot be passed on easily). When the basic step gives $\hat{r} \in (0,0.25)$ then it is possible that the subsequent projection step might give $\tilde{r} \leq 0$. In this case, to then accept the basic step would conflict with the above requirement. However it is unlikely that the projection step will fail if the ratio $\tilde{r}_e$ is first checked for good agreement, so the possibility of returning to the basic step in these circumstances is ignored.

The details of the algorithm which is thus determined are as follows: the k-th iteration is given.

<u>Algorithm (3.3)</u>

(i) given $\underset{\sim}{x}^{(k)}$, $\underset{\sim}{\lambda}^{(k)}$, $\rho^{(k)}$, evaluate $f^{(k)}$, $\underset{\sim}{c}^{(k)}$, $\underset{\sim}{g}^{(k)}$, $A^{(k)}$, $W^{(k)}$ if necessary

(ii) solve the basic step subproblem giving $\underset{\sim}{\delta}^{(k)}$, $\hat{\underset{\sim}{\lambda}}$, $\Delta\psi^{(k)}$

(iii) evaluate $\hat{\underset{\sim}{x}}$, $\hat{\phi}$, $\hat{r}$

if $\hat{r} > 0.75$ goto (ix)

(iv) solve the projection step subproblem giving $\hat{\underset{\sim}{\delta}}$, $\tilde{\underset{\sim}{\lambda}}$, $\Delta\hat{\psi}$, $\tilde{r}_e$

if $\hat{r} < 0.25$ goto (vi)

(v) if $\tilde{r}_e \in [0.9, 1.1]$ set $\rho^{(k+1)} = 2\rho^{(k)}$ and goto (xi)

otherwise goto (x)

(vi) if $\tilde{r}_e \notin [0.75, 1.25]$ goto (vii)

evaluate $\tilde{\underset{\sim}{x}}$, $\tilde{\phi}$, $\tilde{r}$

in computation assign $\underset{\sim}{\delta}^{(k)} := \underset{\sim}{\delta}^{(k)} + \hat{\underset{\sim}{\delta}}$, $\hat{\underset{\sim}{\lambda}} := \tilde{\underset{\sim}{\lambda}}$, $\hat{r} := \tilde{r}$

if $\hat{r} > 0.75$ goto (ix)

if $\hat{r} \in [0.25, 0.75]$ goto (x)

(vii) set $\rho^{(k+1)} \in [0.1, 0.5] ||\underset{\sim}{\delta}^{(k)}||$

if $\hat{r} > 0$ goto (xi)

(viii) set $\underset{\sim}{x}^{(k+1)} = \underset{\sim}{x}^{(k)}$, $\underset{\sim}{\lambda}^{(k+1)} = \underset{\sim}{\lambda}^{(k)}$, end of iteration

(ix) if $\hat{r} > 0.75$ and $||\underset{\sim}{\delta}^{(k)}|| = \rho^{(k)}$ set $\rho^{(k+1)} = 2\rho^{(k)}$ and goto (xi)

if $\hat{r} > 0.9$ and $||\underset{\sim}{\delta}^{(k)}|| = \rho^{(k)}$ set $\rho^{(k+1)} = 4\rho^{(k)}$ and goto (xi)

(x) set $\rho^{(k+1)} = \rho^{(k)}$

(xi) set $\underset{\sim}{x}^{(k+1)} = \hat{\underset{\sim}{x}}$, $\underset{\sim}{\lambda}^{(k+1)} = \hat{\underset{\sim}{\lambda}}$, end of iteration.

In step (vii) $\rho^{(k+1)}$ is set to $\alpha||\underset{\sim}{\delta}^{(k)}||$ where α lies in [0.1, 0.5] . The actual value of α is determined by interpolation to approximately minimize the function $\phi(\underset{\sim}{x}^{(k)} + \alpha\underset{\sim}{\delta}^{(k)})$, ($\phi_\alpha$ say). Exactly what interpolating function to use is not clear however, for instance either a quadratic or a piecewise linear function might be appropriate in different circumstances. Fortunately only a crude estimate of the minimizer is required. The approach is based on attempting to estimate the slope ϕ'_0 (even though this may not exist) and then using quadratic interpolation. Two cases are identified: if the step restriction is active then it is assumed that $\psi^{(k)}(\alpha\underset{\sim}{\delta}^{(k)})$ decreases linearly with α and so $\phi'_0 = -\Delta\psi^{(k)}$. Otherwise it is assumed to vary like a quadratic with a minimizer at $\alpha = 1$, in which case $\phi'_0 = -2\Delta\psi^{(k)}$.

Now ϕ_0 $(= \phi^{(k)})$ and ϕ_1 $(= \hat{\phi})$ are known so fitting a quadratic to ϕ_0, ϕ_1 and ϕ_0' yields the formula $\alpha_{min} = \frac{1}{2}/(1 + (\phi_0 - \phi_1)/\phi_0')$ which gives

$$\begin{aligned} \alpha &= \tfrac{1}{2}/(1 - \hat{r}) \quad \text{if} \quad \|\underset{\sim}{\delta}^{(k)}\| = \rho^{(k)} \\ \alpha &= \tfrac{1}{2}/(1 - \tfrac{1}{2}\hat{r}) \quad \text{if} \quad \|\underset{\sim}{\delta}^{(k)}\| < \rho^{(k)} . \end{aligned} \tag{3.4}$$

These formulae give $\alpha = \frac{1}{2}$ when $\hat{r} = 0$ and α decreases smoothly as $\hat{r}$ decreases which is just what is required. The value determined by (3.4) is truncated to the nearest point in [0.1, 0.5] and then used in algorithm (3.3), step (vii).

Practical experience with this algorithm is encouraging and is described in detail in section 4. Asymptotically, if the step restriction is inactive, then the results of section 2 show that the rate of convergence is second order and that the Maratos effect cannot occur. In the rest of this section the global convergence properties of the algorithm are considered. The proof of global convergence is similar to that in Fletcher (1980a, 1981a) and is outlined here in Theorem 3.1. A discussion of other theoretical results that might be desirable and the extent to which they are likely to be true is given at the end of section 4. The idea of trying other points if the basic step fails is also a feature of the 'watchdog technique' of Chamberlain et. al (1980) as a means of circumventing the Maratos effect. There however it is the basic step that is repeated and a fixed (preset) number of failures is allowed. This calculation is much more expensive than the calculation of the projection step here which only requires a single vector of functions to be evaluated and which can take advantage of parametric programming facilities.

The global convergence proof Theorem 14.3.1 in Fletcher (1981a) (or Theorem 3.1 in Fletcher (1980a) where more detail is given) starts by observing that either

$$\text{(i)} \;\; \rho^{(k)} \to 0 \qquad \text{or} \qquad \text{(ii)} \;\; \inf \rho^{(k)} > 0 .$$

In case (i) there must exist an infinite subsequence such that $\rho^{(k+1)} < \min_{j\le k}\rho^{(j)}$. This implies that $\hat{r} < 0.25$ and $||\underset{\sim}{\delta}^{(k)}|| \to 0$ and any accumulation point $\underset{\sim}{x}^{(k)} \to \underset{\sim}{x}^{\infty}$ of this subsequence is shown to satisfy first order necessary conditions. (Note that for $h^{(k)}$ read $\rho^{(k)}$ and $r^{(k)}$ read $\hat{r}$.) In case (ii) it is possible to find a subsequence for which $\rho^{(k+1)} \ge \rho^{(k)}$ which implies that $\hat{r} \ge 0.25$. A thinner subsequence can be selected for which $\underset{\sim}{\lambda}^{(k)} \to \underset{\sim}{\lambda}^{\infty}$. Any accumulation point of this subsequence is also shown to satisfy first order necessary conditions. The same approach can be used in regard to algorithm (3.3). If $\rho^{(k+1)} < \rho^{(k)}$ then $\hat{r} < 0.25$ must hold whether or not the projection step is used, and so the argument for case (i) goes through as before. In case (ii) when $\rho^{(k+1)} \ge \rho^{(k)}$ then either $\hat{r} \ge 0.25$ or $\tilde{r} \ge 0.25$ must hold. As in Theorem 14.3.1, $\phi^{(1)} - \phi^{\infty} \ge \sum(\phi^{(k+1)} - \phi^{(k)})$ summed over the subsequence, which implies $\phi^{(k+1)} - \phi^{(k)} \to 0$ and hence $\Delta\psi^{(k)} \to 0$ by definition of $\hat{r}$ or $\tilde{r}$. The remainder of the proof follows as in Theorem 14.3.1. Thus the following has been proved.

<u>Theorem 3.1</u> Let $\underset{\sim}{x}^{(k)} \in B \subset \mathbb{R}^n$ where B is compact and let $f, \underset{\sim}{c} \in \mathbb{C}^2$ on B. Then there exists an accumulation point $\underset{\sim}{x}^{\infty}$ of algorithm (3.3) at which first order conditions hold, that is

$$\max_{\underset{\sim}{g}\in\partial\phi^{\infty}} \underset{\sim}{s}^T \underset{\sim}{g} \ge 0 \quad \forall \underset{\sim}{s}. \tag{3.5}$$

□

Two final comments are made. Note that Theorem 3.1 does not claim that every accumulation point of the sequence $\{\underset{\sim}{x}^{(k)}\}$ satisfies (3.5). It is possible that a case (i) subsequence exists with $\hat{r} < 0.25$ and $\underset{\sim}{x}^{\infty}$ is the accumulation point which satisfies (3.5). However for points $\underset{\sim}{x}^{(k)}$ in this subsequence the algorithm does not restrict $\underset{\sim}{\delta}^{(k)}$ other than by requiring that $\phi^{(k+1)} < \phi^{(k)}$. Thus it is not easily possible to discount the existence of other accumulation points which do not satisfy (3.5). Different trust region algorithms require $\underset{\sim}{\delta}^{(k)} = \underset{\sim}{0}$ when $\hat{r} < 0.25$, that is they do not allow the iterates to change until $\rho^{(k)}$ is sufficiently small to force $\hat{r} \geq 0.25$. Sorenson (1980) shows that for this strategy and for smooth unconstrained optimization any limit point becomes stationary. Another comment concerns the observation by Fletcher (1980a, 1981a) that in case (ii), second order necessary conditions hold at $\underset{\sim}{x}^{\infty}$. This is only in regard to the vectors $\underset{\sim}{x}^{\infty}$, $\underset{\sim}{\lambda}^{\infty}$. It has not been shown that $\underset{\sim}{\lambda}^{\infty}$ is a multiplier vector associated with the first order conditions (3.5) and it would be desirable to show that this is the case. However the result is easily proved when $\underset{\sim}{x}^{(k)} \rightarrow \underset{\sim}{x}^{\infty}$ for the main sequence which is the usual practical case.

4. Numerical experiments and discussion

In this section the practical performance of algorithm (3.3) is investigated, and particularly its ability to follow a steep sided curved groove. The numerical experiments are based on using an exact ℓ_1 penalty function to solve nonlinear programming problems and nonlinear systems of equations. Most standard test problems of this nature are not very badly behaved, so two standard problems are modified so that the steepness of the groove sides relative to the slope along the groove can be controlled. In particular the effect of increasing this ratio to very large values is investigated. Firstly the Rosenbrock equations problem is modified to give the 2×2 system of nonlinear equations

$$
\begin{aligned}
c_1(\underset{\sim}{x}) &\triangleq \sigma(x_2 - x_1^2) = 0 \\
c_2(\underset{\sim}{x}) &\triangleq x_1 - 1 \qquad = 0 ,
\end{aligned}
\qquad (4.1)
$$

which is solved by minimizing $\phi(\underset{\sim}{x}) = ||\underset{\sim}{c}(\underset{\sim}{x})||_1$. The parameter σ controls the steepness of the parabolic groove and the usual standard value is $\sigma = 10$. The progress of the basic method (essentially (14.5.6)) and that of algorithm (3.3), both started from $\underset{\sim}{x}^{(1)} = (-1.2, 1)^T$ and $\underset{\sim}{\lambda}^{(1)} = \underset{\sim}{0}$, is shown in table 4.1 .

Table 4.1 Rosenbrock's equations

σ	basic method			Algorithm (3.3)		
	N_D	N_F	ρ	N_D	N_F	ρ
10	12	12	0.25	7	11	
100	193	196	0.00771	7	13	≥ 0.25
1000	~2000	~2000	0.000741	7	13	

N_D and N_F denote the total number of evaluations of all the derivatives and all the functions respectively. N_F is also the total number of calls of the ℓ_1QP subroutine and N_D is the number of calls of that subroutine on which parametric programming cannot be used. It is observed that the trust region radius usually settles down quickly to a fixed value and in the table ρ denotes what this value is. It can be seen for the basic method that the computational effort required to follow the groove goes up in proportion to σ and the typical trust region radius goes down in inverse proportion, so that the time and effort soon becomes excessive as σ is increased. On the other hand, the effort required by algorithm (3.3) is virtually unchanged and the typical trust region radius does not decrease. However, there are special features due to the simple form of (4.1) which imply that the projection step always locates the groove exactly, and so the performance in table 4.1 overstates the abilities of algorithm (3.3).

To get a more realistic picture, a more complicated 3×3 system of nonlinear equations is constructed based on the Helical Valley equations test problem. The equations are

$$\begin{aligned} c_1(\underset{\sim}{x}) &\triangleq \sigma(10\theta - x_3) = 0 \\ c_2(\underset{\sim}{x}) &\triangleq \sigma(x_1^2 + x_2^2 - 1) = 0 \\ c_3(\underset{\sim}{x}) &\triangleq x_3 = 0 \end{aligned} \tag{4.2}$$

where $\theta(x_1, x_2)$ gives the angle in revolutions which the vector (x_1,x_2) makes with the x_1 - axis. Again $\phi(\underset{\sim}{x}) = ||\underset{\sim}{c}(\underset{\sim}{x})||_1$ is minimized. A solution of $c_1(\underset{\sim}{x}) = c_2(\underset{\sim}{x}) = 0$ defines a curved helical groove of radius 1 and pitch 10 and the third equation picks out the point (1, 0, 0) as the solution of (4.2). The relative steepness of the helical groove is controlled by increasing σ as before, and the usual standard value of the parameter is $\sigma = 10$. The standard starting point for this problem is $\underset{\sim}{x}^{(1)} = (-1, 0, 0)^T$ which is not close to the helical groove, and it is possible for the basic method to solve the problem without following the groove, unless σ is very large (see table 4.2). Thus an alternative 'groove start' is considered, that is $\underset{\sim}{x}^{(1)} = (-1, 0, 5)^T$, which is on the helical groove and remote from the solution. From this starting point the iterates must follow the groove the whole time. The performance of the methods is summarized in tables 4.2 and 4.3.

Table 4.2 Helical valley : standard start

σ	basic method			algorithm (3.3)		
	N_D	N_F	ρ	N_D	N_F	ρ
10	10	10	0.5	9	9	1
100	10	10	0.5	10	11	0.5
1000	13	13	0.25	13	18	0.171
10000	~1600	~1600	0.000287	28	58	0.0193

Table 4.3 Helical valley : groove start

σ	basic method			algorithm (3.3)		
	N_D	N_F	ρ	N_D	N_F	ρ
10	23	23	0.25	13	21	0.64
100	~200	~200	0.025	31	60	0.186
1000	~2000	~2000	0.0025	96	193	0.050
10000	~20000	~20000	0.00025	~250	~500	0.020

From table 4.2 it is observed that for $\sigma = 1000$ or less there is little difference in the performance of the two methods because the iterates do not descend into the groove except close to the solution. It is under these circumstances that using the projection step when $\hat{r} \geq 0.25$ would force the iterates into the groove at an earlier stage and consequently slow down the performance of algorithm (3.3). However when $\sigma = 10000$ then both algorithms descend into the groove at an earlier stage, and the superior performance of algorithm (3.3) in following the groove becomes clear. In table 4.3 it is observed that the effort required by the basic method to follow the groove is directly proportional to σ. In fact the iterates soon fall into a very predictable pattern with ρ constant, and the figures for larger values of σ are calculated on the basis of this pattern remaining unchanged, so as not to waste a large amount of computer time. For algorithm (3.3) the iterates also soon settle into a fixed pattern: on each iteration the basic step fails ($\hat{r} < 0$) but the estimate $\tilde{r}_e$ lies in $[0.75, 1.25]$. Consequently the projection step is tried, and succeeds in giving average agreement ($\tilde{r} \in [0.25, 0.75]$). Thus the algorithm does settle on a value of ρ which is close to being as large as possible subject to average agreement being achieved by the combined basic + projection steps, which is the main aim of the heuristics. It can be observed that in contrast to table 4.1, algorithm (3.3) does require more effort to solve the problem as σ is increased. However for the basic method the effort goes up like $\sim\sigma$ which soon becomes unacceptable, whereas for algorithm (3.3) it appears to go up approximately like $\sim\sigma^{\frac{1}{2}}$ which thus enables very badly behaved problems to be tackled in reasonable time.

Algorithm (3.3) has also been applied to a selection of standard nonlinear programming problems for which the exact ℓ_1 penalty function is not badly behaved. The results are summarized in table 4.4, and are very similar to

Table 4.4 Solving exact ℓ_1 penalty functions using algorithm (3.3)

Test problem	N_D	N_F
Freudenstein and Roth	11	12
Powell n = 2	12	20
Powell n = 5	6	8
Colville TP1	3	3
Colville TP2	8	11
Colville TP3	4	4
Dembo 7	13	20 *

* low accuracy : $\phi - \phi_{min} = 2.9_{10}-3$.

those for the basic method (14.5.6) given in Fletcher (1981a) where more details can be found. The low accuracy solution for the Dembo 7 problem is similar to that for the basic method; although the algorithm has performed well globally, the rapid local rate of convergence has not taken over because of the degenerate nature of the constraints at the solution, to computer accuracy. Thus the main conclusions of this work are the following. Whilst the basic QL method + step restriction is adequate for well-behaved problems in which reasonable attention to scaling has been given, use of the second order correction in a projection step extends the method at little extra complexity or housekeeping cost to handle badly behaved problems with steep sided curved grooves. The global convergence and second order rate of convergence proofs for the basic method carry over. However asymptotically the projection step gives a higher order of agreement between actual and predicted function reductions and this enables the Maratos effect to be avoided.

Nonetheless there are still some gaps in the theoretical properties of algorithm (3.3) which it is desirable to eliminate. Firstly the asymptotic results in section 2 concern the unrestricted basic + projection step. It is desirable to show that for $\underset{\sim}{x}^{(k)}$, $\underset{\sim}{\lambda}^{(k)}$ in a neighbourhood of $\underset{\sim}{x}^*$, $\underset{\sim}{\lambda}^*$ at which standard assumptions hold, then there is a threshold value of ρ below which both $\tilde{r}_e$ and $\tilde{r}$ are calculated to lie within [0.75, 1.25]. It then follows that the step restriction becomes inactive for sufficiently large k and the results of section 2 then follow. However I have been unable to establish the asymptotic results of Theorem 2.1 when the step restriction is active. Nonetheless I conjecture that these results can be proved.

Another gap in the theoretical properties is that for smooth unconstrained optimization it can be proved (Fletcher 1980b, Sorensen 1980) that convergence takes place to a stationary point which satisfies both first and second order necessary conditions, whereas it has not been possible to establish this result for algorithm (3.3) in an NDO problem. In fact the result is not true as the following counter example illustrates. Consider the nonlinear programming problem

$$\begin{aligned} &\underset{\underset{\sim}{x}}{\text{minimize}} \quad f(\underset{\sim}{x}) \triangleq \tfrac{1}{2}((x_1 - 1)^2 + x_2^2 - x_3^2) \\ &\text{subject to} \quad c_1(\underset{\sim}{x}) \triangleq - x_1 + x_2^2 = 0 \\ &\qquad\qquad\quad\; c_2(\underset{\sim}{x}) \triangleq - x_1 - x_3^2 = 0 \end{aligned} \tag{4.3}$$

which has solution $\underset{\sim}{x}^* = (\tfrac{1}{2}, \pm 1/\sqrt{2}, 0)^T$ and $\underset{\sim}{\lambda}^* = (\tfrac{1}{2}, 1)$.
Therefore the function

$$\phi(\underset{\sim}{x}) = \tfrac{1}{2} f(\underset{\sim}{x}) + ||\underset{\sim}{c}(\underset{\sim}{x})||_1 \tag{4.4}$$

is an exact ℓ_1 penalty function which shares the same solution. Consider minimizing (4.4) using algorithm (3.3) with $\underset{\sim}{x}^{(1)} = \underset{\sim}{0}$, $\underset{\sim}{\lambda}^{(1)} = \underset{\sim}{0}$ and $\rho^{(1)} = 0.1$. It is easy to see that $\underset{\sim}{\delta}^{(1)} = (0, 0, \pm \rho^{(1)})^T$ and $\hat{\phi} > \phi^{(1)}$. The second order correction also has zero component for $\hat{\delta}_2$ and so does not improve ϕ , since ϕ is minimized by the origin in the (x_1, x_3) plane when $x_2 = 0$. Thus $\underset{\sim}{x}^{(2)} = \underset{\sim}{x}^{(1)}$, $\underset{\sim}{\lambda}^{(2)} = \underset{\sim}{\lambda}^{(1)}$ is taken and the iteration is repeated with $\rho^{(2)} < \rho^{(1)}$, and with the same outcome, and so on. Thus the algorithm fails to make progress. It follows from Theorem 3.1 that first order conditions are satisfied at $\underset{\sim}{x}^{(1)}$, which is true, but $\underset{\sim}{x}^{(1)}$ is not a local minimizer of $\phi(\underset{\sim}{x})$. It is clear that the algorithm fails to account correctly for second order effects. Because $\underset{\sim}{\lambda}^{(1)} = \underset{\sim}{0}$ the algorithm does not include curvature terms for the functions $c_1(\underset{\sim}{x})$ and $c_2(\underset{\sim}{x})$ which in both cases are dominant and of the opposite sign to the corresponding curvature of $f(\underset{\sim}{x})$. Thus every iteration attempts to change x_3 and keep x_2 fixed, which cannot be done, whereas ϕ is best reduced by changing x_2 and keeping x_3 fixed. It is easy to circumvent this example by modifying algorithm (3.3) to update the multipliers when a step fails (that is $\underset{\sim}{\lambda}^{(k+1)} = \hat{\underset{\sim}{\lambda}}$ in step (viii)). However this might not be desirable in general: for example if $\rho^{(1)}$ were large then a very poor estimate of $\underset{\sim}{\lambda}^{(2)}$ might thus be obtained; also it becomes necessary to recompute the matrix $W^{(k)}$ when an iteration fails. It may be better to think along the lines of associating $\underset{\sim}{\lambda}^{(k)}$ more directly with $\underset{\sim}{x}^{(k)}$ (for example (11.1.18) and (11.1.19) in Fletcher, 1981a), rather than taking $\underset{\sim}{\lambda}^{(k)}$ as the multipliers of a previously successful subproblem. This requires some attempt to identify 'locally active discontinuities' as in algorithm 2 of Fletcher (1981b).

References

Chamberlain R M, Lemarechal C, Pedersen H C and Powell M J D. (1980) "The watchdog technique for forcing convergence in algorithms for constrained optimization', University of Cambridge DAMTP Report 80/NA1.

Coleman T F and Conn A R. (1980) "Nonlinear programming via an exact penalty function : Asymptotic analysis", University of Waterloo, Dept of Computer Science Report CS-80-30.

Conn A R and Sinclair J W. (1975) "Quadratic programming via a non-differentiable penalty function", University of Waterloo, Dept of Combinatorics and Optimization Report CORR 75-15.

Fletcher R. (1980a) "A model algorithm for composite NDO problems", University of Kentucky report and in Proc. Workshop on Numerical Techniques in Systems Engineering, to appear in Mathematical Programming Studies.

Fletcher R. (1980b) "Practical methods of optimization, Volume 1, Unconstrained optimization", Wiley, Chichester.

Fletcher R. (1981a) "Practical methods of optimization, Volume 2, Constrained optimization", Wiley, Chichester.

Fletcher R. (1981b) "Numerical experiments with an L_1 exact penalty function method" in "Nonlinear programming 4", eds. O L Mangasarian, R R Meyer and S M Robinson, Academic Press, New York.

Han S P. (1981) "Variable metric methods for minimizing a class of nondifferentiable functions", Math. Prog. 20 pp. 1-13.

Moré J J. (1978) "The Levenberg-Marquardt algorithm : implementation and theory" in "Numerical Analysis, Dundee 1977", ed. G A Watson, Lecture Notes in Mathematics 630, Springer-Verlag, Berlin.

Powell M J D. (1978) "A fast algorithm for nonlinearly constrained optimization calculations" in "Numerical Analysis, Dundee 1977", ed. G A Watson, Lecture Notes in Mathematics 630, Springer-Verlag, Berlin.

Pshenichnyi B N. (1978) "Nonsmooth optimization and nonlinear programming" in "Nonsmooth optimization", eds. C Lemarechal and R Mifflin, IIASA Proceedings 3, Pergamon, Oxford.

Sargent R W H. (1974) "Reduced gradient and projection methods for nonlinear programming" in "Numerical methods for constrained optimization" eds. P E Gill and W Murray, Academic Press, London.

Sorensen D C. (1980) "Newton's method with a trust region modification" Argonne Nat. Lab. Report ANL-80-106.

Wilson R B. (1963) "A simplicial algorithm for concave programming", PhD dissertation, Harvard University Graduate School of Business Administration.

Wolfe M A. (1978) "Extended iterative methods for the solution of operator equations", Numer. Math., 31, pp. 153-174.

AUTOMATIC METHODS FOR HIGHLY OSCILLATORY ORDINARY DIFFERENTIAL EQUATIONS*

C.W. Gear and K.A. Gallivan

Abstract

By a highly oscillatory ODE we mean one whose solution is "nearly periodic." This paper is concerned with the low-cost, automatic detection of oscillatory behavior, the determination of its period, and methods for its subsequent efficient integration. In the first phase, the method for oscillatory problems discussed examines the output of an integrator to determine if the output is nearly periodic. At the point this answer is positive, the second phase is entered and an automatic multirevolutionary method is invoked to integrate a quasi-envelope of the solution. This requires the occasional solution of a nearly periodic initial-value problem over one period by a standard method and the re-determination of its period to provide the approximate derivatives of a quasi-envelope. The major difficulties addressed in this paper are the following: the determination of the point at which multirevolutionary methods are more economic, the automatic detection of stiffness in the multirevolutionary method (which uses a very large step), the calculation of the equivalent Jacobian for the multirevolutionary method (it is a transition matrix of the system over one period), and the calculation of a smooth quasi-envelope.

1. Introduction

The problem of highly oscillatory ODEs has some parallels with that of stiff ODEs: often the solution is not nearly periodic initially, and maybe not even oscillatory, so conventional (nonstiff) methods are best in this transient phase, but after awhile the solution exhibits a nearly periodic behavior and the objective may be to determine the average behavior, the waveform, or its envelope over many millions of cycles. There are some methods that are applicable in the latter nearly periodic phase, for example, Mace and Thomas [6], Graff [5], Graff and Bettis [4], and Petzold [7]. However, these methods cannot be used in the transient phase, so we must detect the onset of nearly periodic behavior. Conversely, a nearly periodic system may cease to be so. This also must be detected so that a switch back to a conventional integrator can be made, just as detection of the termination of stiffness is also desirable, although there it is for the sake of efficiency, not necessity.

---

*Supported in part by the Department of Energy, Contract DE-AC02-76ER02383.A003.

On the other hand, it is important to realize that the problem of highly oscillatory ODEs is, unlike stiff equations, not due to the presence of large eigenvalues. Large eigenvalues may be present and be responsible for the oscillatory behavior but if the system is close to a constant-coefficient linear problem and the oscillation is due to pure imaginary eigenvalues, it suffices to damp these oscillations out. Only in this case is there an analogy with the stiff problem--there are eigencomponents we wish to ignore. However, in the more interesting cases the system is nonlinear and we must track the amplitude and waveshape of the oscillation. (Note that tracking the phase over billions of cycles is an inherently ill-conditioned problem unless the phase is locked to an oscillatory input.)

Methods for nearly periodic problems are generally known as multirevolutionary from their celestial orbit background. The idea of such methods is to calculate, by some conventional integrator, the change in the solution over one orbit. If the period of an orbit is T (for a moment assumed fixed), then a conventional integrator is used to compute the value of

$$D(t,y) = d(t) = y(t + T) - y(t)$$

by integrating the initial value problem $y' = f(t, y)$ over one period T. If we consider the sequence of times $t = mT$, m integral, we have a sequence of values $y(mT)$ which are slowly changing if y is nearly periodic. The conventional integrator allows us to compute the first differences $d(mT)$ of this sequence at any time mT. Under appropriate "smoothness" conditions (whatever that means for a sequence) we can interpolate or extrapolate for values of $d(mT)$ from a subset of all values of d, for example from $d(kqT)$, $k = 1, 2, 3, \ldots$, where q is an integer > 1, and thus estimate $y(mT)$ by integrating only over occasional orbits. The multirevolutionary scheme is summarized in section 2.

In a satellite orbit problem it is fairly easy to define the meaning of "one period." For example, one could use a zero crossing of a particular coordinate, or even a fixed period based on a first order theory. In her thesis, Petzold [7] considered problems for which it is difficult to find physical definitions of the period and examined a method for determining the approximate period by minimizing a function of the form

$$I(t, T) = \int_t^{t+T} \|y(\tau + T) - y(\tau)\| d\tau$$

using a Newton method. The value of T which minimizes $I(t, T)$ is a function of t, and $T(t)$ was said to be the period of the solution. This enabled $d(t) = y(t + T(t)) - y(t)$ to be calculated and multirevolutionary methods to be used. The variable period was handled easily by a change of independent variables to s in which the period is constant, say 1. The equation

$$t(s + 1) - t(s) = T(t(s))$$

was appended to the system

$$z(s + 1) - z(s) = g(s, z)$$

where $z(s) = y(t(s))$ and $g(s, z) = D(t(s), z)$ for integer values of s. (When T is constant, this is the analog of the old device for converting a non-autonomous system to an autonomous system by appending the differential equation $t' = 1$.)

The scheme for period calculation used by Petzold suffers from three drawbacks. The first drawback is that it is fairly expensive, requiring a numerical approximation to the first two derivatives of $I(t, T)$ by quadrature which itself requires the values of $y(\tau)$, $y'(\tau)$, and $y''(\tau)$ over two periods. The second drawback is that a reasonably accurate period estimate is needed for the Newton iteration to converge. Outside the region of convergence of Newton s method a search scheme for a minimum could be used but this would be very expensive because of the computation involved in each quadrature even if all previously computed values could be saved. This makes the approach very unattractive for initial period detection when there is no starting estimate. The third drawback is that minimizing a function subject to several sources of error (including truncation errors in the integration and quadrature, and roundoff errors revealed by considerable cancellation in $\|y(\tau + T) - y(\tau)\|$) is likely to yield a fairly inaccurate answer. Since the value of $d(t) = g(s, z)$ is quite sensitive to small absolute changes in the period T which may be large relative to the period, the function $g(s, z)$ may not appear to be very smooth.

An alternate approach to determination of the period was described in Gear [2]. It also allows for the onset of nearly periodic behavior to be detected and a decision to be made when to switch to multirevolutionary methods. This method can also be used to decide when the solution is no longer nearly periodic. It should be noted that in this case, $T(t)$ and hence $D(t, y)$ and $g(t, y)$ are no longer defined. As Gallivan [1] points out, it is important to use the same technique to decide when to invoke the multirevolutinary methods as used in these methods to control their continued use, or the program may repeatedly switch back and forth. The multirevolutionary and periodic detection/determination techniques will be summarized in sections 3 and 4.

The multirevolutionary method resembles an integration scheme which "integrates" the difference equation $z(s + 1) - z(s) = g(s, z)$ with a large stepsize H. There is a high probability that this will be stiff because $H\partial g/\partial z$ may be large. The calculation of $J = \partial g/\partial z$ is not possible explicitly since it is a transition matrix for the original differential equation. Numerical calculation can be difficult because a numerical perturbation of z may cause g to be undefined. A method for avoiding this problem will be discussed in section 5.

Another source of difficulty arises from the errors in the standard integration method used to compute $g(t, z)$. Current automatic integrators produce an answer which will be within a multiple of the user-specified tolerance ε, but as z is changed, the integrator produces answers which are not smooth functions of z. The effect of this is that a very small tolerance must be used to compute g, or its lack of smoothness causes the outer integrator to reduce its stepsize greatly and to stay with low order. Hence there is interest in constructing automatic integrators whose output is a "smooth" function of its input, that is to say, whose output has the same degree of differentiability with respect to the input as the mathematical problem in the absence of roundoff. The difficulties of this problem and one approach are discussed in section 6.

2. The Quasi-envelope and Multirevolutionary Methods

Suppose, for a moment, that the period $T(t)$ is known. To simplify the discussion we will also take it to be a constant, although neither of these suppositions is necessary. A period T quasi-envelope, $z(t)$, of a function $y(t)$ is any function that agrees with y at the periodic points $t = mT$. We are interested in the case in which the function $y(t)$ is the solution of the initial value problem $y' = f(t, y)$, $y(0) = y_0$, which is nearly periodic with period T, and in a smooth quasi-envelope. For example, if $y(t)$ is periodic, then the best quasi-envelope for our purposes is a constant. The importance of the quasi-envelope is that when we know it we have a low-cost way of computing the solution of the original problem at any point: to find the value of $y(t^*)$ choose the largest integer m such that $mT \leq t^*$ and integrate $y' = f(y, t)$ from $t = mt$, $y(mT) = z(mT)$ to $t = t^*$. If m is very large, this is much less expensive than integrating from the initial conditions at $t = 0$. Hence, from the quasi-envelope and the differential equation we can compute information such as the waveform, amplitude, energy, etc., at any point at a low cost. Note that if the original ODE is autonomous, we can integrate it from any starting point $(t, z(t))$ to determine a waveshape which evolves continuously (and differentiably) in time. The same is approximately true if the ODE is nearly autonomous, that is, $\partial f/\partial t$ is small compared to $1/T$. In these cases it is not necessary to start the integration at a periodic point. We call this the unsynchronized mode. If $\partial f/\partial t$ is large, phase is important. In this case, unless the t-dependence of f determines the period (in which case we say that there is a nearly periodic driving term), the problem is ill-conditioned. Otherwise we can either determine the phase from the driving term in what we call the synchronized mode, or the phase is unimportant.

A multirevolutionary method is a technique for computing a quasi-envelope given a way to compute $z(t + T) - z(t) = d(t)$. For small T this says $z'(t) \cong d(t)/T$.

Hence, it is not surprising that the numerical interpolation for z(t) given a technique for computing d(t)/T is very similar to a numerical integration technique. In the new coordinate system, the basic structure of the program is an outer integrator which solves the equations

$$z(s + 1) - z(s) = g(t(s), z(s))$$
$$t(s + 1) - t(s) = T(t(s))$$

using an outer stepsize H. The method varies the order and stepsize just as an ordinary integrator does. See Petzold [7] for details. It calls a subroutine to evaluate g and T given z and t. This is done by integrating the underlying ordinary differential equation $y' = f(y)$ starting from $y(t) = z$, determining when a period has elapsed and computing $g(t, z) = y(t + T(t)) - y(t)$.

The variable period multirevolutionary integrator is based on a modified Nordsieck scheme. Each component of z is represented by the history vector

$$\underline{a} = [z, Hg, H^2g'/2, H^3g''/6, \ldots, H^kg^{(k-1)}/k!]^T$$

Petzold has shown that in this representation the predictor has the form

$$\underline{a}_{n,(0)} = A\underline{a}_{n-1}$$

where A is the Pascal triangle matrix except for the first row which is

$$[1, 1, \alpha_1(r), \alpha_2(r), \ldots, \alpha_{k-1}(r)]$$

where $r = 1/H$. She also showed that the corrector takes the form

$$\underline{a}_n = \underline{a}_{n,(0)} + \underline{\ell}\omega$$

where ω is chosen so that $\underline{a}_n$ "satisfies" the relation $z(s_n + 1) - z(s_n) = g_n$ and $\underline{\ell}$ is the conventional corrector vector except in the first component which is a function of $r = 1/H$. Petzold gives these functions for generalized Adams methods. (They are polynomials in r.) The corresponding functions for generalized BDF methods are inverse polynomials in r and are given in [2].

3. Periodic Behavior Detection

We have been deliberately imprecise about the meaning of "nearly periodic," and will continue that way with the working definition in our minds of "the type of problem that can be handled efficiently by multirevolutionary methods." We have been equally imprecise about the definition of the "period" of a nearly periodic function. We could use some intuitively reasonable mathematical description, in which case we would have to seek computational algorithms for its approximation. However, the period is most easily defined in terms of the algorithm used to calculate it. It should, of course, yield the exact period for periodic functions and be close for small perturbations of periodic functions. This replaces an analysis of the accuracy of period calculation with an analysis of the efficiency of the multirevolutionary method with respect to different period definitions. This latter may be an easier task.

Petzold's period definition, based on minimizing a norm, is very expensive to apply and cannot be considered as a technique for determining if an arbitrary output of an integrator is nearly periodic. Therefore, we look for alternate definitions of the period. First, note that if the oscillation is due to a periodic driving function, we probably know its period or can examine the system which generates the driving function directly. Hence, we can restrict ourselves to autonomous systems or nearly autonomous systems. A nearly autonomous system can be made autonomous by the substitution $t = v/\varepsilon$ and the additional equation $v' = \varepsilon$. Since v is slowly changing, the enlarged autonomous system may also be nearly periodic.

The solution of an autonomous system is completely determined by the specification of the value of the solution vector y at one time. That is to say, if we identify two times on the solution such that $y(t_1) = y(t_2)$, we know that the solution is periodic with period $t_2 - t_1$. This first suggests determining the period by looking for minimum of $\|y(t_1) - y(t_2)\|$. The cost of this is not particularly low and it requires a clever adaptive program with a lot of heuristics to determine the onset of nearly periodic behavior because we know neither t_1, the value when the behavior first occurs, not $t_2 - t_1$, the period.

A more reliable way of defining the period is to identify certain points on the solution at which a simple characterization is repeated, such as zero crossing. The solution itself may not have zero crossings and, if it consists of a periodic function superimposed on a slowly growing function, there may be difficulty in choosing any value which is crossed periodically. However, its derivative will have periodic sign changes, so we have experimented with a definition of period based on the zero crossings of $c^T y'$ where c is a vector of constants. The program examines the integrator output for positive-going zero crossings of $c^T y'$. (Currently, c is a vector of the weights provided by the user for error norm calculations.) Anything but a simple periodic solution may lead to more than one zero crossing in a single period, so the norm $\|y'(t_1) - y'(t_2)\|$ is also examined, where t_1 and t_2 are a pair of zero crossings. If the norm is small, the possibility of a period is considered. The procedure used is as follows:

1. Identify a positive going sign change in $c^T y'$.

2. Interpolate to find the t value, $t_{current}$, of the zero crossing. Also compute interpolated values of y and y' at the zero crossing.

3. Save these values. (Up to ten prior values are saved in the experimental program.)

4. Compare the current values of y' with each prior value in turn until a small $\|y'_{old} - y'_{current}\|$ is found.

5. Save $T = t_{current} - t_{old}$.

6. Continue to calculate additional periods, T, starting from the latest $t_{current}$ each time. Examine the backward differences of T over several periods. When they are small, indicating a smoothly varying period, consider switching to multirevolutionary methods. Details are given in [1].

The decision on when to switch to multirevolutionary methods is based on estimates of the stepsize H that can be used in the outer integrator. Because the ODE has been integrated over several periods, we have backward differences of $g(t(s_n), z(s_n))$ based on a stepsize in s of $H = 1$. These are used to estimate the order and stepsize that can be used by generalized Adams and BDF methods. The Adams stepsize estimate is also limited to stay in the region of absolute stability and corrector convergence based on a Lipschitz estimate for g. (Initially this is zero, and is updated when two evaluations are done in a PECE Adams steps.) Next, the work factor, H/W, is calculated for each method, where H is the stepsize and W is an estimate of the cost of the multirevolutionary method compared to a non-multirevolutionary method when H is 1. For Adams methods we take $W = 3$ since there are two evaluations of g, each taking one inner integration over a period, and some additional overhead. For BDF, W should be about $3 + n/10$, where n is the number of equations. This is based on assumed Jacobian evaluation every 10 steps, each costing n additional inner integrations over one period.

4. Period Calculation and Stiffness Detection

When using the multirevolutionary method, we need to compute $g(t(s), z)$ and $T(t(s))$ given $z = z_n$ and $t(s) = t_n$. This uses the same technique as the periodic detection except that the vector c must be chosen so that $c^T y'_n \equiv c^T f(t_n, z_n) = 0$. The first step of the inner integrator is executed so that y'_n and y''_n can be calculated and estimated, respectively. Then c is chosen to maximize $c^T y''_n$ subject to $\|c\| = 1$ and $c^T y'_n = 0$. (A single equation requires special treatment here, but it can only be oscillatory if there is an oscillatory driving term.) The inner integrator continues and positive going zero crossings of $c^T y'$ are checked to find one such that $\|y' - y'_n\|$ is small. If a period is not found within 30 of the previously calculated period, it is decreed that the function is no longer nearly periodic from the assigned starting values. This will cause a stepsize reduction in the outer integrator until the periodic detection is successful or the outer stepsize H is so small that the work factor is less than one. This causes a switch back to a conventional method, as would be appropriate if the solution were no longer nearly periodic.

The outer integrator initially uses a generalized Adams method because there is no knowledge of the Lipschitz constant. Two corrector iterations are used, enabling a Lipschitz estimate to be obtained. The step/order selection algorithm is basically that described in the previous section. Whenever the stepsize is estimated, the decision between stiff and nonstiff methods is made based on the current Lipschitz estimate.

5. Jacobian Calculation

When stiff outer methods are used, $\partial y/\partial z$ must be estimated by perturbing z numerically and integrating y over one period. This can cause the value of g to be undefined because of a loss of periodicity. This problem can be circumvented by modifying the inner integration to compute a value of g based on the period calculated for the unperturbed problem using the following algebra:

$$g(t, z) = y(t + T) - z(t)$$

where y is the solution of $y' = f(t, y)$ starting from $(t, z(t))$. Note that T depends on z(t), which we will indicate by T(z). Hence,

$$\frac{\partial g}{\partial z} = \frac{\partial}{\partial z} y(t + T(z)) - I$$

Let T be the value of T(z) at the value of z for which we wish to compute the Jacobian, and let $\partial y(t + T)/\partial z$ be the partial of y with T fixed. Hence,

$$\frac{\partial g}{\partial z} = \frac{\partial y}{\partial z}(t + T) - I + y'(t + T)\frac{\partial T}{\partial z}$$

$\partial y(t + T)/\partial z - I$ can be calculated by numerical differencing of the ouput of the inner integrator over a fixed period, so we need to compute $\partial T/\partial z$. T is defined by the equation $c^T y'(t + T(z) = 0$. Differentiating with respect to z we get

$$\frac{\partial c^T}{\partial z} y'(t + T) + c^T \frac{\partial y'}{\partial z}(t + T) + c^T y''(t + T)\frac{\partial T}{\partial z} = 0$$

or

$$\frac{\partial T}{\partial z} = -\frac{(y')^T \frac{\partial c}{\partial z} + c^T \frac{\partial y'}{\partial z}(t + T)}{c^T y''(t + T)}$$

The quantities $\partial c/\partial z$ and $\partial y'/\partial z$ can be calculated by numerical differencing at the same time that $\partial y(t + T)/\partial z$ is calculated. This allows $\partial g/\partial z$ to be estimated.

6. Smoothness of g

Results reported in [2] and [1] required a very small integration tolerance in the inner integration in relation to the outer integration. If the inner integration tolerance is enlarged, the outer integrator takes smaller steps and has difficulty because the value of g is subject to integration errors which look like random functions of the argument z when an automatic integrator is used. It appears to be worthwhile considering the use of an inner integrator whose output is a smooth function of its input within roundoff error. If the stepsize and order are fixed in an integrator, the output will be as differentiable as the differential equation. Unfortunately, it is not reasonable to fix the stepsize and order a priori. Skelboe [8] uses the same set of stepsizes and orders over successive periods in a code which calculates the periodic steady state, but that strategy does not seem adaptable to this situation.

The alternative is to consider automatic integration techniques whose internal parameters are smooth functions of their inputs. This means that there can be no internal branching to alternate paths such as occur in codes when

(a) A step is rejected for a large error.

(b) The order is changed (by a discrete amount!).

(c) A variable number of iterations of an implicit scheme is used, unless the iteration error is reduced to roundoff level.

(d) Devices such as counters which inhibit changes for some number of steps are used.

All computed functions must be smooth, which means that L_1 and L_∞ norms may not be used. Also, if the output at a particular value of t is obtained by interpolation, great care must be taken with the interpolation formula. Petzold [personal communication] pointed out that most interpolation techniques used in ODE codes are not C_1, and some are not even C_0 because the function is approximated by a different interpolatory polynomial over each interval.

Vu [9] has recently investigated a Runge-Kutta code which has the interesting property that it never rejects a step, it just reduces the stepsize based on the error estimate. It uses the RK starter scheme described in [3] to compute estimates of $y_n^{(p)}$, p = 1, 2, 3, and 4 and an error estimate, and then uses a Taylor series to compute y_{n+1} with a stepsize that can be a smooth function of the error estimate. This can be used to produce a "smooth" automatic integrator and is the subject of current experiments.

7. Conclusion

Results reported in [2] and [1] indicate that some highly oscillatory problems can be integrated very efficiently by these methods. The types of problems that are amenable to these techniques are those with a single oscillation, either due to a driving term or a nonlinear oscillator whose behavior is "stable," that is, whose amplitude and waveform are not sensitive to small perturbations. Essentially this means that the problem is reasonably well posed. The important problem of two or more oscillations at different frequencies cannot be currently handled by these techniques.

References

[1] Gallivan, K.A., Detection and integration of oscillatory differential equations with initial stepsize, order and method selection, Dept. Computer Science Report UIUCDCS-R-80-1045, Univ. Illinois at Urbana-Champaign, M.S. Thesis, 1980.

[2] Gear, C.W., Automatic detection and treatment of oscillatory and/or stiff ordinary differential equations, Dept. Computer Science Report UIUCDCS-R-80-1019, Univ. Illinois at Urbana-Champaign, 1980. To appear in Proceedings of the Bielefeld Conference on Numerical Methods in Computational Chemistry, 1980.

[3] Gear, C.W., Runge-Kutta starters for multistep methods, TOMS 6 (3), September 1980, 263-279.

[4] Graff, O.F. and D.G. Bettis, Modified multirevolution integration methods for satellite orbit computation, Celestial Mechanics 11, 1975, 443-448.

[5] Graff, O.F., Methods of orbit computation with multirevolution steps, Applied Mechanics Research Laboratory Report 1063, Univ. Texas at Austin, 1973.

[6] Mace, D. and L.H. Thomas, An extrapolation method for stepping the calculations of the orbit of an artificial satellite several revolutions ahead at a time, Astronomical Journal 65 (5), June 1960.

[7] Petzold, L.R., An efficient numerical method for highly oscillatory ordinary differential equations, Dept. Computer Science Report UIUCDCS-R-78-933, Univ. Illinois at Urbana-Champaign, Ph.D. Thesis, 1978.

[8] Skelboe, S., Computation of the periodic steady state response of nonlinear networks by extrapolation methods, IEEE Trans. Circuits and Systems CAS-27, (3), 1980, 161-175.

[9] Vu, T., Modified Runge-Kutta methods for solving ODEs, Dept. Computer Science Report UIUCDCS-R-81-1064, Univ. Illinois at Urbana-Champaign, M.S. Thesis, 1981.

Convergence of a Two-Stage Richardson Iterative Procedure for Solving Systems of Linear Equations.

Gene H. Golub* and Michael L. Overton†

0. Introduction

Consider the problem of solving a system of linear equations

$$Ax = b \tag{0.1}$$

by an iterative method, i.e. generating a sequence of approximations $\{x_k\}$ such that $\lim_{k\to\infty} x_k = x$. Frequently it is useful to introduce a splitting

$$A = M - N$$

where systems of the form

$$My = c \tag{0.2}$$

may be solved much more easily than the original system (0.1). When (0.1) arises from the discretization of a partial differential equation, such a splitting is often natural, with M and N corresponding to separate terms in the differential equation. The iterative method used to solve (0.1) can then be "preconditioned" by M, i.e. designed so that each step of this "outer" iteration involves solving a system of the form (0.2). Sometimes it is desirable to solve these systems (0.2) by "inner" iterative procedures. In this paper we consider using the second-order Richardson method for the outer iteration, and show how the rate of convergence of this iteration depends on the accuracy re-

* Computer Science Department, Stanford University, Stanford, California 94305, U.S.A. Supported in part by the United States Department of Energy contract DE-AT03-ER71030 and in part by the National Science Foundation grant MCS-78-11985.

† Computer Science Department, Courant Institute of Mathematical Sciences New York University, 251 Mercer St., New York, NY 10012, U.S.A. Supported in part by the United States Department of Energy contract DE-AC02-76ER03077 and in part by the National Science Foundation grant MCS-81-01924.

quired for the inner iterations. To our knowledge analysis of this particular method has not been attempted previously. See Nichols (1973) for an analysis of more general schemes using inner and outer iterations to solve linear systems. Other related papers on the solution of linear or nonlinear problems by two-stage methods include Gunn (1964), Nicolaides (1975), Pereyra (1967) and Dembo, Eisenstat and Steihaug (1980). See Golub (1962) for a study of round-off error for the Richardson method, which we do not explicitly consider here.

One motivation for our work is that the two-stage method can be effectively used to solve nonsymmetric systems where M is symmetric positive definite and N is skew-symmetric. In this case the symmetric/skew-symmetric splitting $A = M-N$ is used to precondition the outer iteration, and a symmetric splitting $M = M_1-M_2$ can be used to precondition each inner iteration, using a direct method to solve a system of the form $M_1 y = c$ at each step of each inner iteration. (See Manteuffel (1977) for alternative approaches for nonsymmetric systems.) Numerical results from applying the Richardson method to such nonsymmetric systems are given in Section 3. We also present numerical results using the conjugate gradient method for the outer iteration. Analysis of the latter procedure would be very interesting, but this seems more difficult.

We use $\|\cdot\|$ to denote the Euclidean vector and matrix norm, defined by

$$\|x\| = (x^T x)^{1/2} \quad , \quad \|B\| = \max_{\|x\| \neq 0} \frac{\|Bx\|}{\|x\|} \; .$$

1. Symmetric positive definite systems.

Let us first assume that both A and M are symmetric and positive definite. Consider the following iterative method.

Method 1.

Choose positive scalar parameters δ, α and ω, with $0 \leq \delta < 1$. Choose initial vectors x_0 and x_1. For $k = 1,2,\ldots$, define

$$x_{k+1} = x_{k-1} + \omega(\alpha z_k + x_k - x_{k-1}) \tag{1.1}$$

where

$$Mz_k = r_k + q_k \; , \tag{1.2}$$

$$r_k = b - Ax_k \; ,$$

and

$$\|q_k\| \leq \delta \|r_k\|. \tag{1.3}$$

If $\delta = 0$ then Method 1 reduces to the second-order Richardson method. The conditions on the linear operator and on the parameters α and ω under which convergence is guaranteed are well known in this case (see Golub and Varga (1961)). When $\delta > 0$ the meaning of (1.2) and (1.3) is that the system

$$Mz = r_k \tag{1.4}$$

is being "solved" by an inner iterative procedure, which is terminated when the residual norm $\|q_k\|$ has been sufficiently reduced. If the outer (main) iteration is converging to the solution of (0.1), the associated residuals $\{r_k\}$ are converging to zero, and hence $z = 0$ is a reasonable starting point for each inner iteration. Equation (1.3) then specifies that each inner iteration must reduce its associated residual norm by a factor of δ. Note that the early inner iterations solve (1.4) with low absolute accuracy, while the later systems are solved with high accuracy. There is no restriction on the type of method used for the inner iterations.

1.1 Convergence analysis

Let us now analyse the convergence of Method 1. Let the error at each step of the outer iteration be

$$e_k = x - x_k,$$

where x is the solution of (0.1). Note that $r_k = Ae_k$ for all k. For $k = 1,2,\ldots,$ we have:

$$\begin{aligned} e_{k+1} &= e_{k-1} - \omega(\alpha z_k + e_{k-1} - e_k) \\ &= \omega K e_k + (1-\omega)e_{k-1} + p_k \end{aligned} \tag{1.5}$$

where

$$p_k = -\omega\alpha M^{-1} q_k$$

and

$$K = I - \alpha M^{-1} A. \tag{1.6}$$

Now K is similar to a symmetric matrix, and we can define its eigenvector decomposition by

$$K = M^{-1/2}(I-\alpha M^{-1/2}AM^{-1/2})M^{1/2} = M^{-1/2}V\Sigma V^T M^{1/2} \tag{1.7}$$

where Σ is a diagonal matrix of eigenvalues $\{\sigma_j\}$, $j = 1,\ldots,n$, and V is orthogonal. Let

$$\hat{e}_k = V^T M^{1/2} e_k \text{ and } \hat{p}_k = V^T M^{1/2} p_k = -\omega\alpha V^T M^{-1/2} q_k \,. \tag{1.8}$$

Then from (1.5) we obtain the diagonalized system of difference equations:

$$\hat{e}_{k+1} = \omega\Sigma\hat{e}_k + (1-\omega)\hat{e}_{k-1} + \hat{p}_k \;,\; k = 1,2,\ldots \,. \tag{1.9}$$

At this point we state a lemma which can be proved using the standard theory of difference equations.

<u>Lemma 1</u>. Consider the inhomogeneous difference equation in ξ_k:

$$\xi_{k+1} = \beta\xi_k + \gamma\xi_{k-1} + \eta_k,\; k = 1,2,\ldots \,. \tag{1.10}$$

The solution to (1.10) is given by

$$\xi_k = \theta_k\,\xi_1 + \gamma\theta_{k-1}\,\xi_0 + \sum_{\ell=1}^{k-1} \theta_{k-\ell}\eta_\ell$$

where

$$\theta_k = \begin{cases} \dfrac{\lambda_1^k - \lambda_2^k}{\lambda_1 - \lambda_2} & \text{if } \lambda_1 \neq \lambda_2 \\ k\lambda_1^{k-1} & \text{if } \lambda_1 = \lambda_2 \end{cases}$$

and where λ_1 and λ_2 are the roots of the characteristic polynomical

$$\lambda^2 - \beta\lambda - \gamma = 0.$$

(Note that if the coefficients of (1.10) are real, λ_1 and λ_2 may be complex but θ_k is real). □

It follows from Lemma 1 that the solution to (1.7) is

$$\hat{e}_k = S_k\hat{e}_1 + (1-\omega)S_{k-1}\hat{e}_0 + \sum_{\ell=1}^{k-1} S_{k-\ell}\hat{p}_\ell,\quad k = 1,2,\ldots, \tag{1.11}$$

where S_k is a diagonal matrix with j^{th} diagonal element

$$(S_k)_{jj} = \begin{cases} \dfrac{\lambda_{1,j}^{\,k} - \lambda_{2,j}^{\,k}}{\lambda_{1,j} - \lambda_{2,j}} & \text{if } \lambda_{1,j} \neq \lambda_{2,j} \\[2ex] k\lambda_{1,j}^{\,k-1} & \text{if } \lambda_{1,j} = \lambda_{2,j} \end{cases} \tag{1.12}$$

and where $\lambda_{1,j}$ and $\lambda_{2,j}$ are the roots of

$$\lambda^2 - \omega\sigma_j\lambda + (\omega-1) = 0, \quad j = 1,\ldots,n. \tag{1.13}$$

Now let

$$\rho = \max_{1\le j\le n} (\max(|\lambda_{1,j}|, |\lambda_{2,j}|)). \tag{1.14}$$

It can easily be shown (by induction on k) that

$$\|S_k\| \equiv \max_{1\le j\le n} |(S_k)_{jj}| \le k\rho^{k-1} .$$

Note also from (1.13) that the product of the roots $\lambda_{1,j}\,\lambda_{2,j}$ is equal to $\omega-1$, so $\rho^2 \ge |\omega-1|$. We therefore have from (1.11) that

$$\|\hat{e}_k\| \le k\rho^{k-1}\|\hat{e}_1\| + (k-1)\rho^k\|\hat{e}_0\| + \sum_{\ell=1}^{k-1} (k-\ell)\rho^{k-\ell-1}\, \|\hat{p}_\ell\|, \; k = 1,2,\ldots \tag{1.15}$$

At this point we need a lemma relating $\|\hat{p}_k\|$ to $\|\hat{e}_k\|$.

<u>Lemma 2</u>. Assume that (1.3) holds. Then

$$\|\hat{p}_k\| \le \varepsilon\|\hat{e}_k\| , \quad k = 1,2,\ldots$$

where $\varepsilon = \delta\,\omega\alpha\|M^{-1/2}\|\,\|AM^{-1/2}\|$. (1.16)

Proof. By (1.8) and (1.3) we have:

$$\|\hat{p}_k\| \le \omega\alpha\|M^{-1/2}\|\,\|q_k\|$$

$$\le \delta\,\omega\alpha\|M^{-1/2}\|\,\|r_k\|$$

$$\leq \delta\omega\alpha \| M^{-1/2} \| \| AM^{-1/2} VV^T M^{1/2} e_k \| ,$$

from which the result follows. □

Continuing with the error analysis, we can substitute $\varepsilon\|\hat{e}_\ell\|$ for $\|\hat{p}_\ell\|$ in (1.15), and rewrite it to obtain

$$\|\hat{e}_k\| - \varepsilon \sum_{\ell=1}^{k-1} (k-\ell)\rho^{k-\ell-1} \|\hat{e}_\ell\| \leq k\rho^{k-1} \|\hat{e}_1\| + (k-1)\rho^k \|\hat{e}_0\|, \qquad k = 1,2,\ldots,m, \quad (1.17)$$

where m is introduced to indicate the last computed iterate x_m. This system of m linear inequalities can be written using matrix notation as

$$(I - \varepsilon L) \begin{bmatrix} \|\hat{e}_1\| \\ \\ \|\hat{e}_m\| \end{bmatrix} \leq s$$

where the non-negative strictly lower triangular matrix L and the vector s are defined accordingly. Now $L^m = 0$ so

$$(I-\varepsilon L)^{-1} = I + \varepsilon L + \varepsilon^2 L^2 + \ldots + \varepsilon^{m-1} L^{m-1}.$$

Thus $(I-\varepsilon L)^{-1}$ is also a non-negative matrix and hence

$$\begin{bmatrix} \|\hat{e}_1\| \\ \\ \|\hat{e}_m\| \end{bmatrix} \leq (I-\varepsilon L)^{-1} s . \quad (1.18)$$

Let us define $t = [\tau_1,\ldots,\tau_m]^T = (I-\varepsilon L)^{-1}s$. By definition, τ_k satisfies

$$\tau_k = k\rho^{k-1} \|\hat{e}_1\| + (k-1)\rho^k \|\hat{e}_0\| + \varepsilon \sum_{\ell=1}^{k-1} (k-\ell)\rho^{k-\ell-1} \tau_\ell, \qquad k = 1,2,\ldots,m.$$

The first equation defines $\tau_1 = \|\hat{e}_1\|$. For convenience, we now define $\tau_0 = -\|\hat{e}_0\|$, and it then follows from Lemma 1 that τ_k is the solution to a difference equation of the form (1.10) with inhomogeneous term $\varepsilon\tau_k$ and characteristic polynomial with a double root, i.e.

$$\tau_{k+1} = 2\rho\tau_k - \rho^2\tau_{k-1} + \varepsilon\tau_k, \quad k = 1,2,\dots .$$

This, however, can be viewed as the <u>homogeneous</u> equation

$$\tau_{k+1} = (2\rho+\varepsilon)\tau_k - \rho^2\tau_{k-1}$$

and hence, using Lemma 1 again, has solution

$$\tau_k = \frac{\nu_1^k - \nu_2^k}{\nu_1 - \nu_2}\tau_1 - \rho^2 \frac{\nu_1^{k-1} - \nu_2^{k-1}}{\nu_1 - \nu_2} \tau_0 , \tag{1.19}$$

where ν_1 and ν_2 are the roots of

$$\lambda^2 - (2\rho+\varepsilon)\lambda + \rho^2 = 0.$$

If we define $\tilde{\rho} = \max(\nu_1,\nu_2)$ we have

$$\tilde{\rho} = \rho + \frac{\varepsilon}{2} + (\rho\varepsilon + \frac{\varepsilon^2}{4})^{1/2} \tag{1.20}$$

$$= \rho + \rho^{1/2}\varepsilon^{1/2} + \frac{\varepsilon}{2} + O(\varepsilon^{3/2}), \text{ if } \varepsilon << 1.$$

Using (1.18) and (1.19) we now have a condition which guarantees the convergence of the error norms $\{\|\hat{e}_k\|\}$ to zero, namely $\tilde{\rho} < 1$. More specifically, we have the following result:

<u>Theorem 1</u>. The error norm $\|\hat{e}_k\|$ associated with the kth iterate of Method 1 is bounded by

$$\|\hat{e}_k\| \le k\tilde{\rho}^{k-1} \|\hat{e}_1\| + (k-1)\tilde{\rho}^k \|\hat{e}_0\| ,$$

where $\tilde{\rho} = \tilde{\rho}(\delta,\alpha,\omega)$ is given by (1.20),(1.16),(1.14),(1.13) and (1.7). □
Note that if $\delta = \varepsilon = 0$, we have $\tilde{\rho} = \rho$ and Theorem 1 reduces to the standard convergence property of the second-order Richardson method (see Golub and Varga (1961)). We also note that it is possible to choose the first iterate x_1 in such a way that the error bound of Theorem 1 is somewhat reduced.

1.2 Optimal parameters.

There are three free parameters in Method 1, namely δ, α and ω. Let us first suppose that $\delta = 0$, so that the inner systems (1.4) are solved exactly. In this case the optimal choices of α and ω, i.e. those that minimize ρ in (1.14), may be described using Young's theory of successive overrelaxation. Let

$$|\sigma_\ell| = \max_{1\le j\le n} |\sigma_j|,$$

where $\{\sigma_j\}$ are the eigenvalues of $K = I-\alpha M^{-1}A$, as before. Given any α, the optimal choice of ω is obtained by choosing $\lambda_{1,\ell} = \lambda_{2,\ell} = \frac{1}{2}\omega\sigma_\ell$ in (1.13), i.e.

$$\omega = \frac{2}{1 + \sqrt{1-\sigma_\ell^2}} . \tag{1.21}$$

It can be verified that this choice results in

$$\rho = |\lambda_{1,\ell}| = |\lambda_{1,j}| = |\lambda_{2,j}| = \tfrac{1}{2}\omega|\sigma_\ell| = \sqrt{\omega-1}$$

for all j, $j = 1,\ldots,n$. Note that (1.21) implies $1 \le \omega < 2$. See Varga (1962,p.109) or Young (1971,p.171) for the proof that (1.21) is optimal, and see Golub and Varga (1961) for a description of the connection between the Richardson and successive overrelaxation methods.

Since $\rho = \frac{1}{2}\omega|\sigma_\ell|$, and ω satisfies (1.21), it is clear that α should be chosen to minimize $|\sigma_\ell|$, the spectral radius of K. Let μ_{max} and μ_{min} be respectively the maximum and minimum eigenvalues of $M^{-1}A$. Clearly $|\sigma_\ell|$ is minimized by specifying

$$1 - \alpha\mu_{min} = -(1-\alpha\mu_{max}),$$

i.e.

$$\alpha = \frac{2}{\mu_{max} + \mu_{min}} \tag{1.22}$$

and

$$|\sigma_\ell| = \frac{\mu_{max} - \mu_{min}}{\mu_{max} + \mu_{min}} < 1 .$$

When we permit $\delta > 0$ the optimal choice of parameters is more complicated. Our aim is to minimize the amount of work needed to obtain a solution of prescribed accuracy. A reasonable measure of the total amount of work required for generating $x_1,\ldots,x_m$ is

$$W = \sum_{k=1}^{m} m_k$$

where m_k is the number of iterates required in the kth inner iteration to achieve (1.2) and (1.3). This measure is particularly appropriate when each step of every <u>inner</u> iteration involves solving another system of equations by a direct method. Now let γ be a measure of relative accuracy prescribed for the outer iteration, e.g. specifying $\|\hat{e}_m\| \leq \gamma\|\hat{e}_0\|$. Roughly speaking, Theorem 1 says that the error is reduced by approximately a factor of $\tilde{\rho}$ at each step of the outer iteration (this neglects $\|\hat{e}_1\|$ and a factor of $(k+1)/k$). With this assumption the number of iterates required for the outer iteration is approximately

$$m \approx \frac{-\log \gamma}{-\log \tilde{\rho}(\delta,\alpha,\omega)} \quad .$$

Now let us suppose that the iterative method used for the inner iteration is such that the associated residual is reduced by about a factor of θ at each step. Assuming a starting point $z = 0$ for each inner iteration we have

$$m_k \approx \frac{-\log \delta}{-\log \theta} \quad , \quad k = 1,\ldots,m.$$

Under these assumptions the total amount of work W is about

$$W \approx \frac{-\log \gamma}{-\log \theta} \cdot \frac{-\log \delta}{-\log \tilde{\rho}(\delta,\alpha,\omega)} \quad .$$

Now γ and θ are fixed so a reasonable goal in choosing the parameters δ, α and ω is to minimize

$$\overline{W} = \frac{-\log \delta}{-\log \tilde{\rho}(\delta,\alpha,\omega)} \quad .$$

Naturally this is difficult to do, but the formula gives some insight. As $\delta \to 0$, $\overline{W} \to \infty$, indicating that solving the inner iterations too accurately is very expensive. On the other hand, if δ is allowed large enough so that $\tilde{\rho} \to 1$, we have $\overline{W} \to \infty$, indicating that the outer iteration is not convergent. The optimal value of δ, given α and ω, is somewhere in between these extremes.

The optimal values of α and ω are equally complicated, since $\tilde{\rho}$ depends on them through both ρ and ε, but it seems likely that (1.21) and (1.22) would be good choices. One reasonable way to choose the parameters in practice is to make several numerical experiments. Fortunately, once an adequate choice of parameters is determined for a particular problem, whole classes of related problems can usually be solved efficiently with the same parameter values. Another interesting and promising idea is to try to choose δ dynamically, as the outer iteration proceeds.

2. Symmetric/skew-symmetric splittings.

Method 1 can be used to solve (0.1) if A is nonsymmetric but with a positive definite symmetric part. We define the splitting $A = M-N$ by

$$M = \frac{1}{2}(A^T+A), \quad N = \frac{1}{2}(A^T-A),$$

where M and N are respectively called the symmetric and skew-symmetric parts of A, and M is positive definite. The matrix K is given by

$$K = I-\alpha M^{-1}A = (1-\alpha)I + \alpha M^{-1}N.$$

Now $M^{-1}N$ is similar to the skew-symmetric matrix $M^{-1/2}NM^{-1/2}$, and thus its eigenvalues, say $\{i\kappa_j\}$, $j = 1,\dots,n$, are imaginary. The eigenvalues of K are thus given by

$$\sigma_j = (1-\alpha) + i\alpha\,\kappa_j \,. \tag{2.1}$$

Let us now investigate the choice of optimal parameters when $\delta = 0$. Young (1971,p.196) gives formulae for the optimal choice of ω and the resulting value of ρ in the case when the eigenvalues of K are complex but confined to the rectangle defined by the real and imaginary parts of the largest eigenvalue in modulus, a property which (2.1) satisfies. It can be verified that ρ is minimized when α in (2.1) is chosen by

$$\alpha = 1. \tag{2.2}$$

With this choice the eigenvalues $\{\sigma_j\}$ are imaginary, and the optimal choice of ω reduces to (1.21). Notice that now $\lambda_{1,\ell} = \lambda_{2,\ell}$ is imaginary and $0 < \omega \leq 1$.

Having motivated the choice (2.2), we now permit $\delta > 0$ and consider the convergence of Method 1 using the symmetric/skew-symmetric splitting, with $\alpha = 1$. All of the error analysis, including Theorem 1, applies to this case, provided only that we replace V^T by V^H in (1.7), where V is now a complex unitary matrix (and Σ is imaginary).

3. Numerical Experiments

Consider the differential equation:

$$-\Delta u + (au)_x + au_x + (bu)_y + bu_y + cu = f \tag{3.1}$$

where u,a,b and f are functions defined on the unit square: $0 \leq x \leq 1$, $0 \leq y \leq 1$. Approximating (3.1) by second-order finite differences on a grid with t interior points in each direction results in an $t^2 \times t^2$ linear system to be solved:

$$(M-N)u_t = d. \tag{3.2}$$

Here u_t is the solution to the discretized problem, $N = -N^T$ is a skew-symmetric matrix corresponding to the first-order terms in (3.1), $M = M^T = M_1 - M_2$ is a symmetric matrix, with M_1 the negative discrete Laplacian operator and M_2 a diagonal matrix corresponding to the zero-order term, and d corresponds to f, modified to account for the boundary conditions. We consider the particular differential equation whose solution is $u(x,y) = \exp(x^2+y^2)$, with coefficients $a(x,y) = b(x,y) = 5\exp(x^2+y^2)$, $c(x,y) = 10\exp(3x^2+3y^2)$, and with nonhomogeneous Dirichlet conditions imposed on the boundary of the square.

Table 1 gives the result of solving (3.2) by Method 1. Note that the symmetric/skew-symmetric splitting was used for the outer iteration, as discussed in Section 2. The parameters were $\alpha = 1$ and ω given by (1.21) (see below). The initial vector x_0 was set to zero, and x_1 was obtained by: $x_1 = z_0 + x_0$; $Mz_0 = r_0 + q_0$; $\|q_0\| \leq \delta \|r_0\|$. The termination criterion was $\|r_m\| \leq 10^{-8}$. Each inner iteration used a (preconditioned) conjugate gradient method, starting with the zero vector, to approximately solve a system of the form $Mz = r$, in the sense of (1.2) and (1.3). The symmetric splitting $M = M_1 - M_2$ was exploited so that each step of the inner iteration used a fast direct method to solve a system of the form $M_1\bar{z} = \bar{r}$, i.e. the Poisson equation.

In addition to the Richardson results, Table 1 also gives the results of experiments using a (preconditioned) conjugate gradient method for the outer iteration to solve (3.2). See Concus and Golub (1976) and Widlund (1978) for a discussion of this method, which exploits the symmetric/skew-symmetric splitting. The inner iterations were carried out exactly as above, being terminated by (1.2) and (1.3). The conjugate gradient method was also run with the inner iterations carried out to machine precision, in order to obtain σ_ℓ from the resulting tri-diagonal Lanczos matrix. These runs, made solely to obtain the "optimal" ω defined by (1.21), are not included in the table.

Table 1.

t	δ	METHOD 1. (RICHARDSON).		CONJUGATE GRADIENT	
		m	W	m	W
15	.001	43	332	36	266
	.01	44	261	38	218
	.1	44	169	52	182
	.2	47	137	-	>500
	.5	62	117	-	"
	.6	71	115	-	"
	.7	75	109	-	"
	.8	147	147	-	"
31	.001	57	446	39	279
	.01	46	270	42	235
	.1	48	180	71	237
	.2	52	152	-	>500
	.5	68	132	-	"
	.6	78	127	-	"
	.7	86	132	-	"
	.8	162	162	-	"

Note: t is the number of mesh points in each direction, δ defines (1.4), m is the number of outer iterates and W is the total number of inner iterates (number of calls to the fast Poisson solver).

The results of Table 1 are graphed in Figure 1. Notice that the optimal choice of δ, given ω and α, is quite well-determined by the curves, and has a value which can be close to one. Using a near optimal value of δ is far more efficient than carrying out the inner iterations to full accuracy. It is interesting to note that the Richardson method (with near optimal ω, in practice unknown) is less efficient than the

Figure 1.

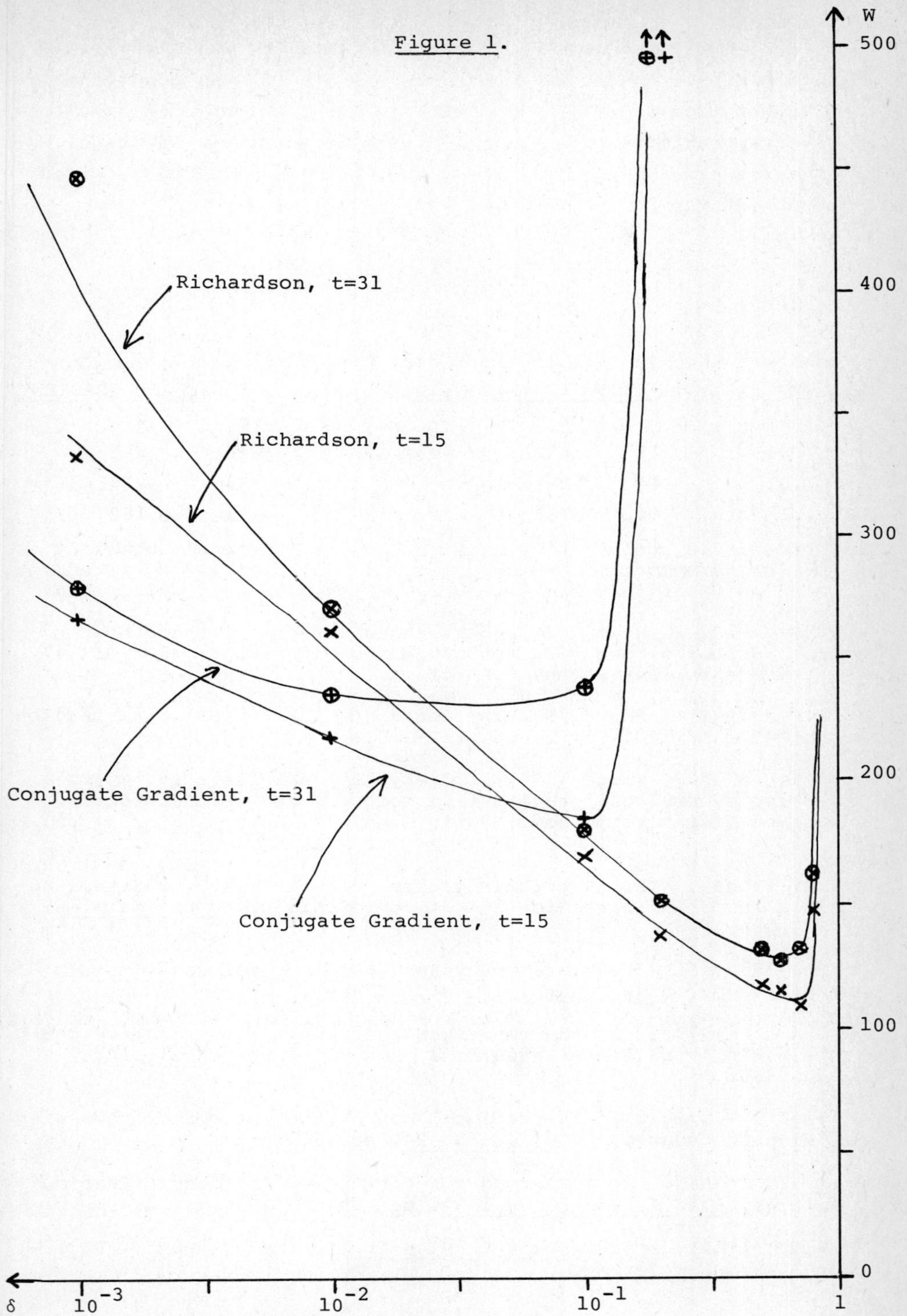
W
500
400
300
200
100
0
Richardson, t=31
Richardson, t=15
Conjugate Gradient, t=31
Conjugate Gradient, t=15
δ
10^{-3}
10^{-2}
10^{-1}
1

conjugate gradient method when the inner iterations are carried out quite accurately, but more efficient when a near optimal δ is used.

The numerical results were obtained using a VAX-11/780 at the Courant Mathematics and Computing Laboratory. Double precision arithmetic, i.e. approximately 16 decimal digits of accuracy, was used. More extensive experiments have been made, but they will be reported in a subsequent paper.

Acknowledgements.

The authors would like to thank Olof Widlund, Don Rose and Nancy Nichols for several helpful conversations and suggestions.

References

Concus, P. and G.H. Golub (1976). A generalized conjugate gradient method for nonsymmetric systems of equations, Computer Science Dept. Report STAN-CS-76-646, Stanford University, Stanford, California.

Dembo, R.S., S.C. Eisenstat and T. Steihaug (1980). Inexact Newton methods, School of Organization and Management Report (Series #47), Yale University, New Haven, Connecticut.

Golub, G.H. (1962). Bounds for the round-off errors in the Richardson second-order method, BIT 2, pp. 212-223.

Golub, G.H. and R.S. Varga (1961). Chebyshev semi-iterative methods, successive overrelaxation iterative methods, and second order Richardson iterative methods, Parts I and II, Numer. Math. 3, pp. 147-168.

Gunn, J.E. (1964). The numerical solution of $\nabla \cdot a \nabla u = f$ by a semi-explicit alternating-direction iterative technique, Numer. Math. 6, pp. 181-184.

Manteuffel, T.A. (1977). The Tchebyshev iteration for nonsymmetric linear systems, Numer. Math. 28, pp. 307-327.

Nichols, N.K. (1973). On the convergence of two-stage iterative processes for solving linear equations, SIAM J. Numer. Anal. 10, pp. 460-469.

Nicolaides, R.A. (1975). On the local convergence of certain two step iterative procedures, Numer. Math. 24, pp. 95-101.

Pereyra, V. (1967). Accelerating the convergence of discretization algorithms, SIAM J. Numer. Anal. 4, pp. 508-533.

Varga, R.S. (1962). Matrix Iterative Analysis, Prentice-Hall, Englewood Cliffs, New Jersey.

Widlund, O. (1978). A Lanczos method for a class of nonsymmetric systems of linear equations, SIAM J. Numer. Anal. 15, pp. 801-812.

Young, D.M. (1971). Iterative Solution of Large Linear Systems, Academic Press, New York and London.

CURVED KNOT LINES AND SURFACES WITH RULED SEGMENTS

J.G. Hayes

1. INTRODUCTION

Sometimes in practical problems it is required to approximate discrete bivariate data with a surface which contains a ruled segment. This type of surface occurs, for example, in some ship hull designs. In such a case a transverse section of one of the symmetric halves of the hull consists of a pair of curves joined by a straight line, with discontinuity of slope at each of the two joins. If the position of the section moves along the vessel, the two points of discontinuity describe a pair of curves ("knuckle lines"), between which is the ruled segment. Ideally, the ruled segment should be developable, since the design objective is to simplify the construction of the hull from flat sheets. In practice, this is achieved to a sufficient approximation by suitable choice of the knuckle lines at the design stage.

It is with surfaces of this kind that we shall be concerned in this paper. Denoting the dependent variable by z and the independent variables by x and y, it will be assumed that each of the two curves of discontinuity bounding the ruled segment can be represented by a pair of equations of the form $z = f(y)$, $x = g(y)$, where f and g are both single-valued. It will be further assumed that $g(y)$ is known, but $f(y)$ may or may not be known. In the ship problem, f and g are indeed single-valued, and both are usually known in the sense that they can be obtained by curve-fitting given data before fitting the surface data.

The representation of the surface will depend on the use of bicubic splines with curved knot-lines, introduced in [7]. That paper is not widely available, and so the required portions of it will first be summarized in Sections 2 and 3. It may be noted that the above paper deals with the situation where x, y and z are all expressed as bicubic spline functions of a pair of parameters, as well as with that where z is expressed as a function of x and y. In the present paper, however, we shall be concerned only with the latter situation. The introduction of straight line segments into the spline curve representation will be treated in Section 4 and the extension to ruled segments in Section 5. Finally, the problem of fitting the surface representation to arbitrarily-scattered data points will be discussed in Section 6 in a number of different circumstances.

For simplicity we shall deal mainly with surfaces having just one ruled segment, but the results apply equally to any number of such segments, as will be indicated. Similarly, we shall use only cubic and bicubic splines, though the results can be readily extended to splines of any degree.

2. DEFINITIONS

A cubic spline $s(x)$ with interior knots $\lambda_3, \lambda_4, \ldots, \lambda_{h-2}$ (where $\lambda_3 < \lambda_4 < \ldots < \lambda_{h-2}$), defined on the interval $[a, b]$, is a function with the following properties.

(i) In each of the intervals

$a \le x \le \lambda_3$; $\lambda_i \le x \le \lambda_{i+1}$ $(i = 3, 4, \ldots, h-3)$; $\lambda_{h-2} \le x \le b$,

$s(x)$ is a cubic polynomial.

(ii) $s(x)$ and its first and second derivatives are continuous for all x in $[a, b]$.

(The numbering of the knots here is a little different from usual but, for cubic splines, helps to clarify some of the results obtained.) If required, the continuity of the spline can be reduced by allowing knots to coincide. In particular, three knots coinciding at a point allows a discontinuity in first derivative at the point.

The spline will be represented in terms of normalized B-splines [4]. To define the full set of B-splines required, it is necessary to choose four extra, artificial, knots at or outside each end of the interval $[a, b]$. We shall choose them to coincide at the appropriate end point, so that we have

$$\lambda_{-1} = \lambda_0 = \lambda_1 = \lambda_2 = a, \text{ and } \lambda_{h-1} = \lambda_h = \lambda_{h+1} = \lambda_{h+2} = b. \tag{2.1}$$

The normalized cubic B-spline $N_i(x;\lambda)$, for $i = 1, 2, \ldots, h$, is a cubic spline with knots $\lambda_{-1}, \lambda_0, \ldots, \lambda_{h+2}$ (collectively denoted by λ) which is non-zero only in the interval $\lambda_{i-2} < x < \lambda_{i+2}$. This defines $N_i(x;\lambda)$ uniquely except for a constant factor. This factor is such as to give the set of normalized B-splines the property that

$$\sum_{i=1}^{h} N_i(x;\lambda) = 1 \tag{2.2}$$

for all x in $[a, b]$. When required, the values of the B-splines at any given value of x are best computed by the method due to Cox [1] and De Boor [4].

We note that the non-zero segment of any particular B-spline spans five knots (or, equivalently, four consecutive intervals between knots, some of which will be of zero length if any of the knots coincide). Because of the labelling chosen for the knots, the central one of these five knots has the same suffix as the B-spline itself. Equally, the four B-splines which are non-zero in the knot interval $[\lambda_t, \lambda_{t+1}]$, for any $t = 2, 3, \ldots, h-1$, are N_{t-1}, N_t, N_{t+1} and N_{t+2}. It is these symmetries between the B-spline suffices and the knot suffices which are helpful in interpreting some of the results we shall obtain.

With the full set of h normalized B-splines, any cubic spline with interior knots $\lambda_3, \lambda_4, \ldots, \lambda_{h-2}$ can be uniquely expressed over the range [a, b] in the form (see [3])

$$s(x) = \sum_{i=1}^{h} \gamma_i N_i(x_i;\lambda), \tag{2.3}$$

where the γ_i are constants.

A bicubic spline s(x, y) is defined on a rectangular region R, given by $a \le x \le b$, $c \le y \le d$. R is divided into rectangular panels (Figure 1) by two sets of lines $x = \lambda_3, \lambda_4, \ldots, \lambda_{h-2}$ and $y = \mu_3, \mu_4, \ldots, \mu_{k-2}$, where

$$a < \lambda_3 < \lambda_4 \ldots < \lambda_{h-2} < b, \text{ and } c < \mu_3 < \mu_4 < \ldots < \mu_{k-2} < d.$$

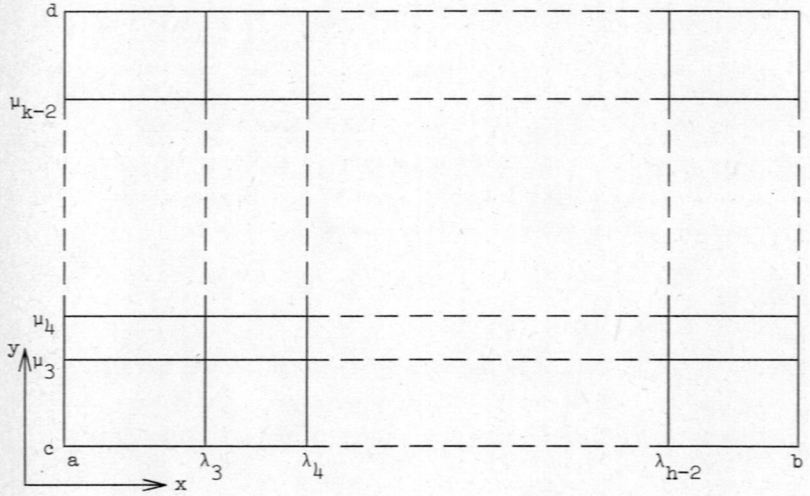

Figure 1. The rectangle R on which the bicubic spline is defined

Then a bicubic spline s(x, y) on R with respect to the given mesh is a function which is a bicubic polynomial of the form

$$s(x, y) = \sum_{r=o}^{3} \sum_{t=o}^{3} \alpha_{rt} x^r y^t$$

in each panel separately, and which has continuity up to the second derivative everywhere in R. Precisely, all partial derivatives $\partial^{p+q}s/\partial x^p \partial y^q$ for $o \le p \le 2$ and $o \le q \le 2$ are continuous.

Any bicubic spline with the given mesh can be uniquely expressed over the rectangle R in the form

$$s(x, y) = \sum_{i=1}^{h} \sum_{j=1}^{k} \gamma_{ij} N_i(x;\lambda) N_j(y;\mu), \tag{2.4}$$

where, corresponding to λ above, the knot set μ for the B-splines in y is composed of $\mu_{-1}, \mu_0, \ldots, \mu_{k+2}$, in which

$$\mu_{-1} = \mu_0 = \mu_1 = \mu_2 = c, \text{ and } \mu_{k-1} = \mu_k = \mu_{k+1} = \mu_{k+2} = d.$$

As in one variable, knots can be allowed to coincide: for example, if $\lambda_3 = \lambda_4 = \lambda_5$, $\partial s(x, y)/\partial x$ will in general be discontinuous across the whole length of the line $x = \lambda_3$ in R.

Re-writing (2.4) in the form

$$s(x, y) = \sum_{i=1}^{h} \left\{ \sum_{j=1}^{k} \gamma_{ij} N_j(y;\mu) \right\} N_i(x;\lambda), \tag{2.5}$$

we note that the bicubic spline can be viewed as the surface generated by a cubic spline of the form (2.3) as its coefficients are allowed to vary as a function of y, i.e. as a cubic spline in y. We shall several times be taking this view of a surface, as being generated by its section at a given value of y as that value varies.

3. CURVED KNOT LINES

The adaptability of the bicubic spline for approximating a variety of surface shapes is a little disappointing when compared with the cubic spline for approximating curves. For example, the cubic spline can usually approximate quite efficiently a curve consisting of a sharp peak with a long region of small curvature on each side of it. It is simply necessary to take a number of closely-spaced knots in the region of the peak and a few widely-spaced knots each side. In a corresponding surface case, however, where we have a sharp ridge running across the surface, such as between the pair of broken curves in Figure 2, the situation is not so satisfactory. If the function s(x, y) in (2.5) is to represent such a surface, it must in particular reduce to a spline curve with a sharp peak between the points P and Q (Figure 2) when y is equated to a constant, K say. Thus the knot-set λ must contain a number of close knots in that part of the range of x spanned by the line PQ. This applies for all values of K in the range of interest and so the knot-set must contain closely-spaced knots over the whole range of x spanned by the ridge. Similarly the knots μ_j must be closely spaced over the whole range of y. Thus the representation of the surface is excessively bulky, and there may be other undesirable consequences. For example, if s(x, y) is to be fitted by, say, least squares to a set of discrete data points [6], there may simply be more coefficients γ_{ij} in (2.4) than there are data to determine them. Even if that is not the case, the

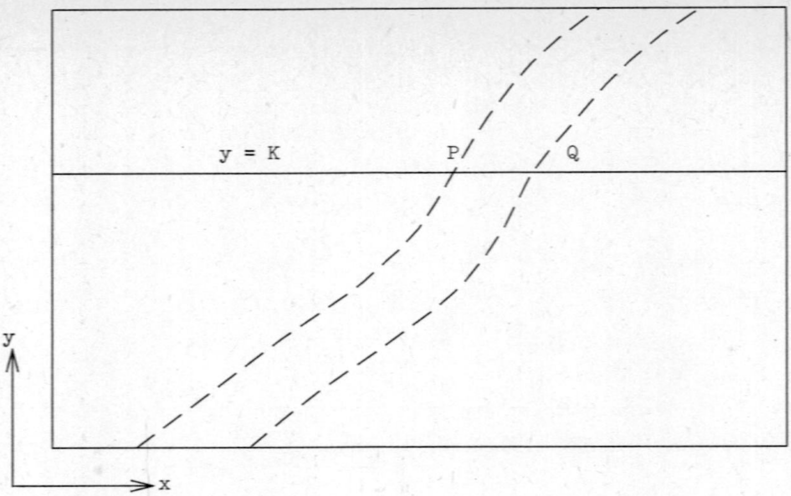

Figure 2. Boundaries of curved ridge

fineness of the knot mesh is likely to produce unwanted fluctuations in the flatter regions away from the ridge, where a much coarser mesh would be appropriate.

Clearly, the surface could be approximated more satisfactorily if the sections of the bicubic spline for different y values could have different knot sets. Thus we are led to the idea of curved knot lines, that is, of allowing the knots λ_i to be functions of y, say $\lambda_i(y)$. Then, denoting the $\lambda_i(y)$ collectively by $\lambda(y)$, we have, instead of (2.4), the generalized form

$$s(x, y) = \sum_{i=1}^{h} \sum_{j=1}^{k} \gamma_{ij} N_i(x;\lambda(y)) N_j(y;\mu). \qquad (3.1)$$

We note (i) that, though (3.1) reduces to a cubic spline for y = constant, it does not do so for x = constant because of the presence of $\lambda(y)$ in the B-splines in x; (ii) that, if second derivative continuity of the surface is to be maintained, each of the functions $\lambda_i(y)$ must have second derivative continuity; (iii) the $\lambda_i(y)$ must be single valued and must satisfy $\lambda_i(y) \le \lambda_{i+1}(y)$ for all i and all y in [c, d].

Then to deal with the problem of the sharp ridge, we simply choose, to form the knot set $\lambda(y)$, a number of curved knot-lines between the broken curves of Figure 2, more or less following contours of the ridge, and a few others elsewhere. After that, it may well be necessary to choose only a small number of knots μ_j for the variable y. Thus we shall have achieved a substantial reduction in the number of knots required, and be in a situation much more comparable with the satisfactory situation in the one-variable case.

Of course, we can go the stage further and allow curved knot-lines for the variable y also, replacing μ in (3.1) by $\mu(x)$. We then have a surface form whose sections both for constant x and constant y are no longer splines. However, in this paper we shall be concerned only with the form (3.1) as it stands.

One result of curved knot lines is that, by choosing three of them to coincide, we can produce a surface which has a slope discontinuity across a curve which traverses the whole surface. More interesting, however, is the fact that we can provide a surface for which the curve of discontinuity does not traverse the whole surface, the discontinuity gradually fading out. Figure 3 shows an example of a set of knot-lines which can produce such a surface. This set consists of three knot-lines which coincide over part of their length but are separate elsewhere. Thus if we consider the plane section at y = K of a surface of the form (3.1) with these knot-lines, it will be a cubic spline with three coincident knots and will therefore in general have a discontinuity in slope. The plane section at y = L, on the other

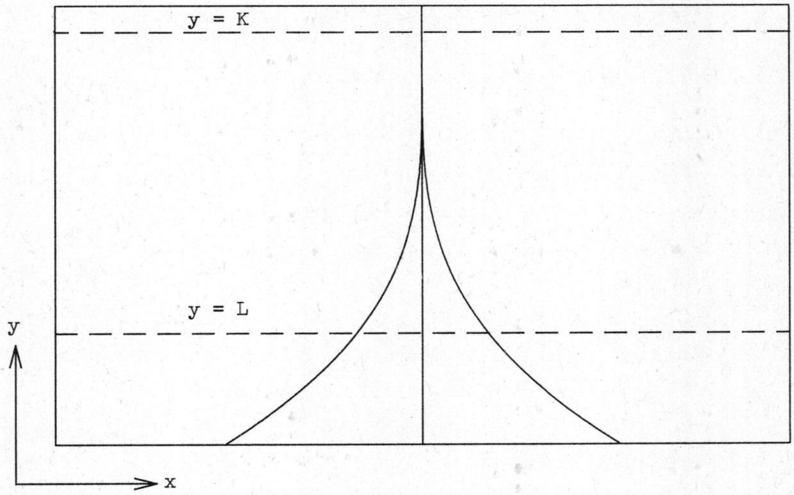

Figure 3. A knot-set producing a fading discontinuity

hand, will be a cubic spline with distinct knots, and will therefore have normal continuity. The latter shape will change smoothly into the former as the knot-lines come together. A simple example of a bicubic spline surface with these knot-lines is shown in Figure 4.

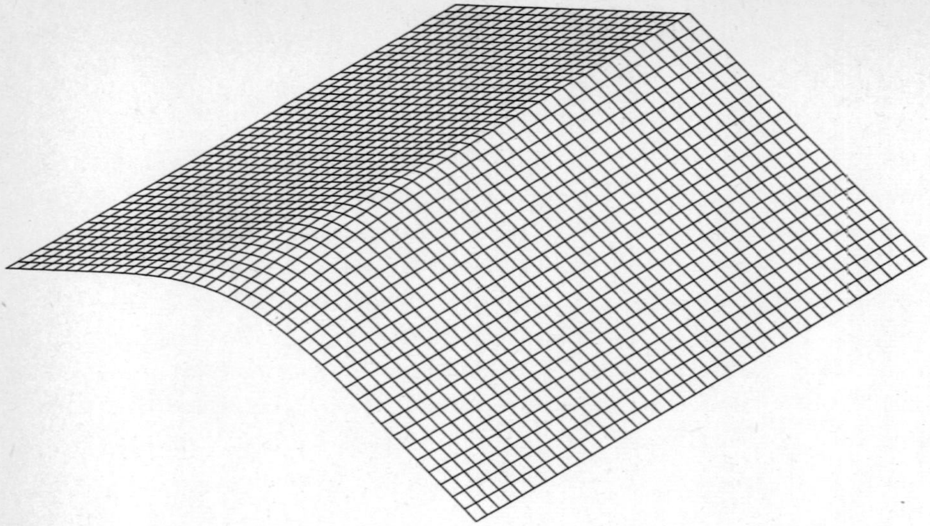

Figure 4. A surface containing a fading discontinuity in slope

4. CURVE WITH STRAIGHT SEGMENTS

We are concerned in this section with the cubic spline

$$s(x) = \sum_{i=1}^{h} \gamma_i N_i(x;\lambda), \tag{4.1}$$

which is the representation (2.3) reproduced here for convenience. To introduce straight lines into the spline representation, we have Greville's result in [5] that

$$\sum_{i=1}^{h} \tfrac{1}{3}(\lambda_{i-1}+\lambda_i+\lambda_{i+1})N_i(x;\lambda) = x, \tag{4.2}$$

for all x in [a, b]. We see that, given the full knot set

$$\lambda_{-1} \quad \lambda_0 \quad \lambda_1 \quad \lambda_2 \quad \cdots \quad \lambda_{h-1} \quad \lambda_h \quad \lambda_{h+1} \quad \lambda_{h+2},$$

the coefficients in (4.2) are obtained by deleting the first and last knot and taking the running mean of adjacent triples in the knot set remaining; also that the coefficient of N_i is the mean of the central three of the five knots spanned by the non-zero segment of N_i.

Now, as well as (4.2), we have the identity (2.2). Taking a linear combination of these two identities gives

$$\sum_{i=1}^{h} \left[\tfrac{1}{3}\, p(\lambda_{i-1}+\lambda_i+\lambda_{i+1})+q\right] N_i(x;\lambda) = px+q, \tag{4.3}$$

for all x in [a, b], which provides the means to represent any straight line over that interval. Writing

$$z_i = p\lambda_i + q, \quad i = 0, 1, \ldots, h + 1, \tag{4.4}$$

it follows from (4.3) that the straight line $z = px + q$ can be represented in the form

$$z = \sum_{i=1}^{h} \tfrac{1}{3}(z_{i-1}+z_i+z_{i+1})N_i(x;\lambda), \tag{4.5}$$

where we note that the z_i are the ordinates of the line $z = px+q$ at the λ_i. Equally, we can see that, if we are given points (λ_i, z_i), for $i = 0, 1, \ldots, h+1$, lying on a straight line, equation (4.5) reproduces that line.

As an example, let us consider the special case where there are no interior knots. Then $h = 4$, and we have only end-knots

$$\lambda_{-1} = \lambda_0 = \lambda_1 = \lambda_2 = a \text{ and } \lambda_3 = \lambda_4 = \lambda_5 = \lambda_6 = b.$$

If we are given ordinate values f_1 at $x = a$ and f_2 at $x = b$, the z values on the line joining the two points (a, f_1) and (b, f_2) are

i :	0	1	2	3	4	5
z_i:	f_1	f_1	f_1	f_2	f_2	f_2.

Taking averages of successive triples, we have from (4.5) that the equation of the straight line joining the two points is

$$z = f_1N_1(x;\lambda) + \tfrac{1}{3}(2f_1+f_2)N_2(x;\lambda) + \tfrac{1}{3}(f_1+2f_2)N_3(x;\lambda) + f_2N_4(x;\lambda). \tag{4.6}$$

We note that the coefficients are in arithmetic progression. The four B-splines have the forms shown in Figure 5.

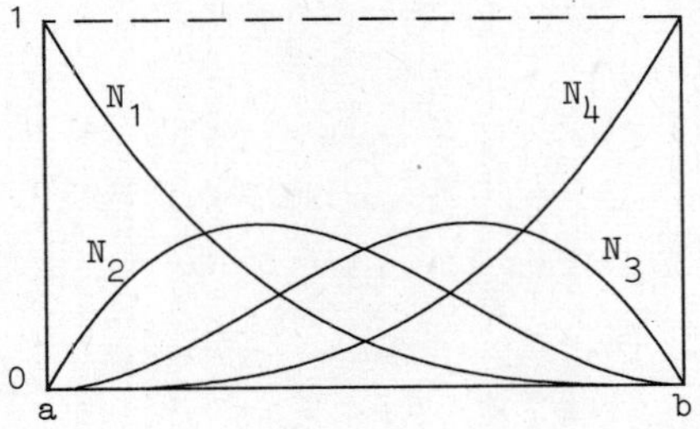

Figure 5. The normalized cubic B-splines when there are no interior knots.

Next, returning to the general case, we note that, since only four B-splines are non-zero in any particular knot interval, for a cubic spline to be a straight line in that interval it is sufficient that only the four corresponding coefficients be determined as in (4.5). Specifically, in the interval $[\lambda_t, \lambda_{t+1}]$, say, the four non-zero B-splines are N_{t-1}, N_t, N_{t+1} and N_{t+2}, and so the spline will be linear in this interval if (λ_i, z_i) for $i = t-2, t-1, \ldots, t+3$ lie on a straight line. In particular, when $\lambda_{t-2} = \lambda_{t-1} = \lambda_t = x_1$, say, and $\lambda_{t+1} = \lambda_{t+2} = \lambda_{t+3} = x_2$, say, then it can be readily seen, as for (4.6), that the spline (4.1) will have a straight line segment joining the two points (x_1, f_1) and (x_2, f_2) provided that

$$\begin{aligned} \gamma_{t-1} &= f_1, \\ \gamma_t &= \tfrac{1}{3}(2f_1+f_2) \\ \gamma_{t+1} &= \tfrac{1}{3}(f_1+2f_2) \\ \gamma_{t+2} &= f_2, \end{aligned} \tag{4.7}$$

regardless of the values of the other coefficients and of the positions of the other knots. The spline will have slope discontinuities at x_1 and x_2, in general. The four relevant B-splines are displayed in Figure 6. In the interval (x_1, x_2)

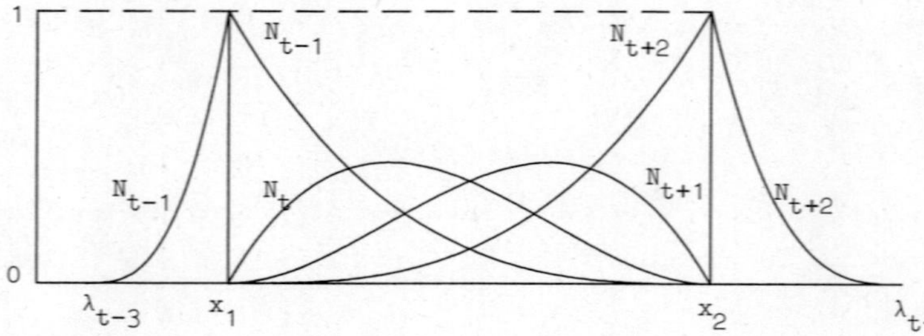

Figure 6. The four normalized cubic B-splines which are non-zero in the interval $[x_1, x_2]$ when x_1 and x_2 are triple knots.

itself, the B-splines have exactly the same form as the B-splines in Figure 5, provided the two interval lengths x_2-x_1 and $b-a$ are the same. (This is a particular instance of the fact that, whether or not there are coincident knots, only B-spline segments lying outside the interval $[\lambda_t, \lambda_{t+1}]$ are affected by λ_{t-3} and λ_{t+4}).

If, further, we choose the next three knots after λ_{t+3} to coincide at x_3, and take

$$\gamma_{t+3} = \tfrac{1}{3}(2f_2+f_3), \quad \gamma_{t+4} = \tfrac{1}{3}(f_2+2f_3), \quad \gamma_{t+5} = f_3,$$

for some f_3, then the spline is straight also in the interval $[x_2, x_3]$, in fact joining the points (x_2, f_2) and (x_3, f_3). In other words, with triple knots the conditions for straightness in two adjacent intervals do not interfere with one another. This gives us the means (not very useful!) to represent by a cubic spline a function which is continuous and piecewise linear over the whole range [a, b].

More useful, however, is the fact that it shows how we can represent a curve containing any number of straight segments interspersed in any manner with curved segments.

The result (4.7) can alternatively be derived in the manner now outlined. For the spline (4.1) to be straight in $[\lambda_t, \lambda_{t+1}]$, it follows from (4.3) and the uniqueness of spline representations that, for some p and q, we must have

$$\gamma_i = \tfrac{1}{3}p(\lambda_{i-1}+\lambda_i+\lambda_{i+1}) + q, \quad i = t-1, \dots t+2. \tag{4.8}$$

Eliminating p and q from these four equations, we can obtain

$$\frac{\gamma_t - \gamma_{t-1}}{\lambda_{t+1} - \lambda_{t-2}} = \frac{\gamma_{t+1} - \gamma_t}{\lambda_{t+2} - \lambda_{t-1}} = \frac{\gamma_{t+2} - \gamma_{t+1}}{\lambda_{t+3} - \lambda_t} \tag{4.9}$$

as necessary conditions for the spline to be straight in $[\lambda_t, \lambda_{t+1}]$. It is also easy to show that they are sufficient conditions. Then, in the case $\lambda_{t-2} = \lambda_{t-1} = \lambda_t = x_1$, $\lambda_{t+1} = \lambda_{t+2} = \lambda_{t+3} = x_2$, all the denominators in (4.9) are equal and so the conditions reduce to the four γ's being in arithmetic progression. Now from Figure 6, we see that N_{t-1} takes unit value at x_1 and all the other B-splines are zero there. Thus γ_{t-1} must equal f_1 if the spline is to pass through the point (x_1, f_1). Similarly γ_{t+2} must equal f_2 if the spline is to pass through (x_2, f_2). Equations (4.7) then follow immediately as the necessary and sufficient conditions for the spline to have a straight segment joining the two points.

We note that this alternative derivation more explicitly separates the conditions for linearity from those for passing through the two points.

5. SURFACE WITH RULED SEGMENTS

We have shown that, with $\lambda_{t-2} = \lambda_{t-1} = \lambda_t = x_1$ and $\lambda_{t+1} = \lambda_{t+2} = \lambda_{t+3} = x_2$, the spline curve

$$z = \sum_{i=1}^{t-2} \gamma_i N_i(x;\lambda) + f_1 N_{t-1}(x;\lambda) + \tfrac{1}{3}(2f_1+f_2)N_t(x;\lambda)$$

$$+ \tfrac{1}{3}(f_1+2f_2)N_{t+1}(x;\lambda) + f_2 N_{t+2}(x;\lambda) + \sum_{i=t+3}^{h} \gamma_i N_i(x;\lambda) \tag{5.1}$$

has a straight segment joining the points (x_1, f_1) and (x_2, f_2). For our purpose, the important feature to note about (5.1) is that it contains this segment no matter what are the values of f_1, f_2, all the λ's and all the γ's. In the general case, we saw in (4.3) that, for the spline to have a straight segment, the coefficients of the B-splines had to be specific functions of the λ's. The simple but important step from (4.3) to (4.5) turns attention from the λ's, which are abscissae values, to the z's, which are ordinate values and more useful for our purpose. But of course the 4 coefficients relevant to a particular interval still depend on the λ's. A sufficient condition for the spline to be a straight line in an interval is that the appropriate six points (λ_i, z_i) lie on the straight line. However, when the six points coalesce into two triplets, we are left with only two distinct points, which must lie on a straight line whatever the z values. (Equivalently, we saw that the λ's dropped out of the conditions (4.9)).

It follows at once that if f_1, f_2, the γ's and the λ's in (5.1) are functions of a third variable, y, we have the equation of a surface whose sections at y = constant contain a straight line segment. In other words, the surface contains a ruled segment which joins the two space curves $x = x_1(y)$, $z = f_1(y)$ and $x = x_2(y)$, $z = f_2(y)$ by straight lines in planes orthogonal to the y axis.

In particular, if f_1, f_2 and the γ's are cubic splines with a common knot set, the surface is a bicubic spline with curved knot lines for x and straight knot lines for y. Thus the surface

$$z = \sum_{i=1}^{h} \sum_{j=1}^{k} \gamma_{ij} N_i(x;\lambda(y)) N_j(y;\mu), \tag{5.2}$$

in which

$$\lambda_{-1}(y) \leq \ldots \leq \lambda_{t-2}(y) = \lambda_{t-1}(y) = \lambda_t(y) < \lambda_{t+1}(y) = \lambda_{t+2}(y) = \lambda_{t+3}(y) \leq \ldots \leq \lambda_{h+2}(y)$$

for all values of y in its range [c, d], and in which

$$\begin{aligned} \gamma_{tj} &= \tfrac{1}{3}(2\gamma_{t-1,j} + \gamma_{t+2,j}) \\ \gamma_{t+1,j} &= \tfrac{1}{3}(\gamma_{t-1,j} + 2\gamma_{t+2,j}) \end{aligned} \tag{5.3}$$

for j = 1, 2, ..., k, contains a ruled segment joining the space curve

$$x = \lambda_t(y),\ z = \sum_{j=1}^{k} \gamma_{t-1,j} N_j(y;\mu) \text{ to the curve } x = \lambda_{t+1}(y),\ z = \sum_{j=1}^{k} \gamma_{t+2,j} N_j(y;\mu).$$

It is therefore the type of surface discussed in our Introduction.

Figure 7 shows a bicubic spline of the above kind. The graphical algorithm used, which evaluates the surface on an equi-spaced rectangular grid in the (x, y) plane and then joins adjacent points by straight lines, does not cope too well with curved discontinuities. Thus it was necessary to devise a surface with quite severe discontinuities for their presence to show up well, but even so the two curves on which they lie appear rather fuzzy. Also the sharp peak which should appear near the upper end of the right-hand boundary section has been cut off through joining points each side of it. The lower curve of discontinuity meets the same section, of course, at the lower end of the long straight segment of the section, but is not detectable in the nearby region.

Analogously to the fading discontinuity discussed in relation to Figure 3, we can also produce a surface in which the ruled segment does not carry across the whole

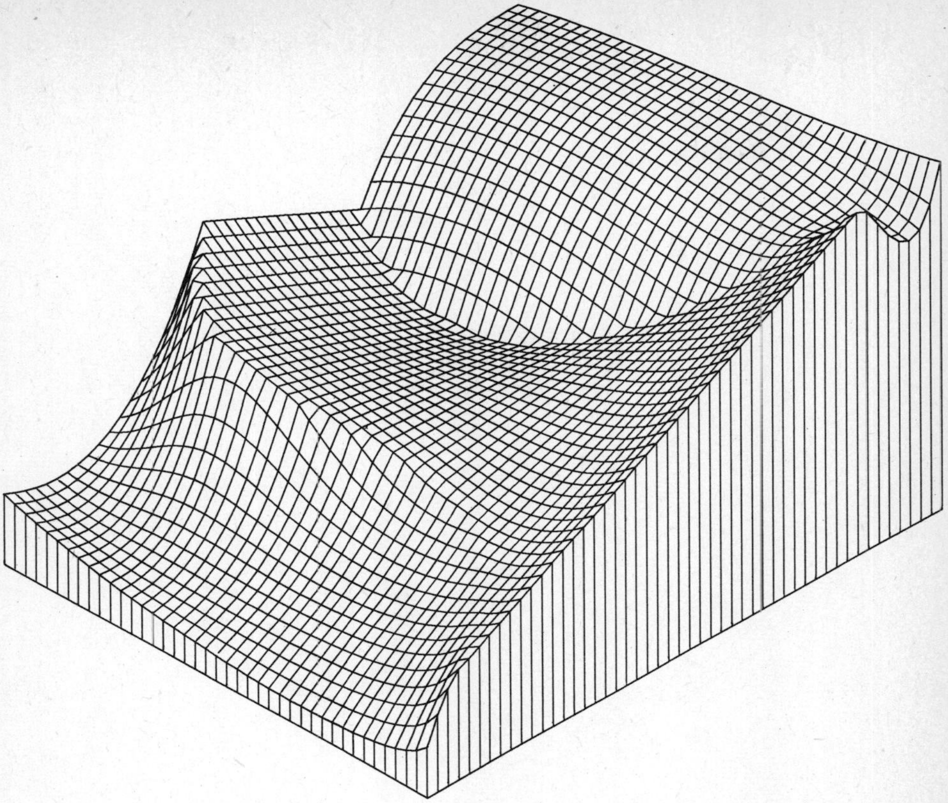

Figure 7. A bicubic spline surface containing a ruled segment

surface, simply by making the pair of triple knot lines coincide for part of the y range. (Figure 8 shows an example of such a surface.) This means that some of the sections for y = constant have six coincident knots, whereas a cubic spline is customarily defined to have no more than four (four allows a jump discontinuity in the value of the spline). However, this causes little difficulty in practice. With a cubic spline in the form (2.3), if a pair of triple knots are brought together, the range over which two of the B-splines are non-zero degenerates to have zero length, but these B-splines can simply be taken to be zero for all x and the value of their coefficients is immaterial. The spline is exactly the same as if there were only four coincident knots. Correspondingly, with the type of surface proposed, the two B-splines $N_t(x;\lambda(y))$ and $N_{t+1}(x;\lambda(y))$ are zero for all x when y is in the range of coincidence of the six knot lines, denoted, say, by $\alpha \le y \le \beta$. If, in addition, there are B-splines in y which are zero for all y outside this range, then clearly there will be terms in (5.2) which are zero everywhere. Precisely, these are the terms with i=t or t+1 and j such that $\mu_{j-2} \ge \alpha$ and $\mu_{j+2} \le \beta$. They exist only when

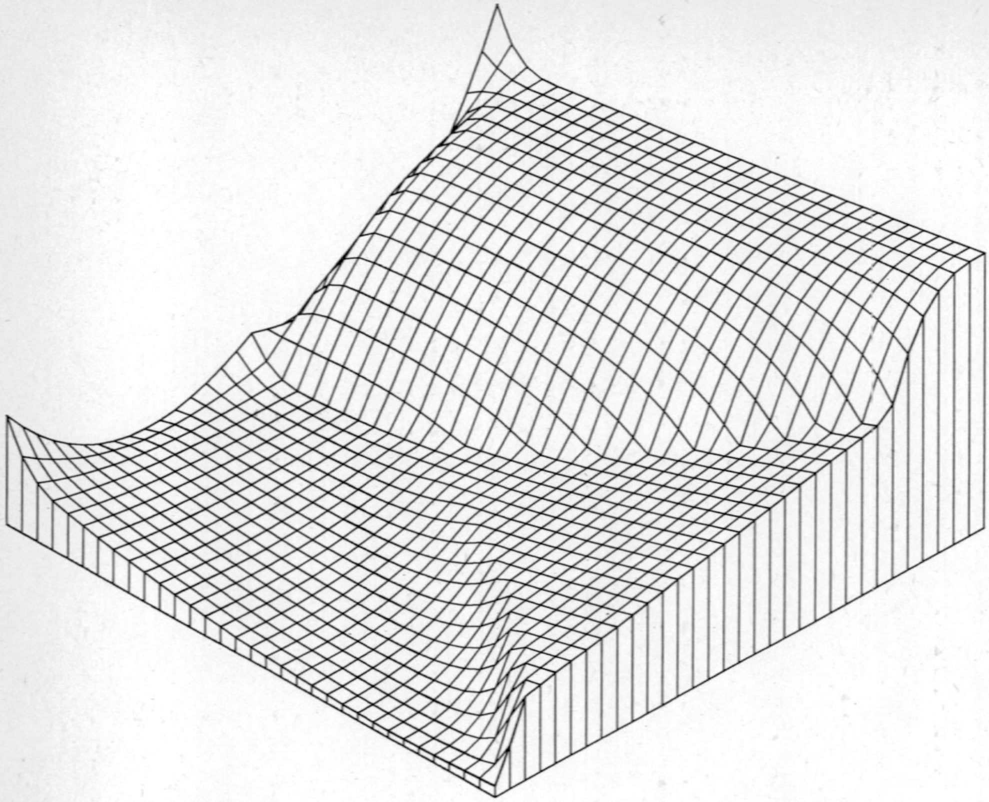

Figure 8. A ruled segment which does not traverse the whole surface.

there are at least five knots in the closed interval $[\alpha, \beta]$. Their coefficients can be given arbitrary values. However, to avoid a jump discontinuity in the value of the surface across the six coincident knot lines, we must ensure that

$$\gamma_{t-1,j} = \gamma_{t+2,j} \tag{5.4}$$

for all j such that $N_j(y;\mu)$ is non-zero somewhere in $[\alpha, \beta]$, that is, for all j such that $\mu_{j-2} < \beta$, $\mu_{j+2} > \alpha$ and $\mu_{j-2} < \mu_{j+2}$.

6. FITTING BY LEAST SQUARES

In discussing the fitting of these surfaces to discrete data, it will be assumed that all the knot lines are known beforehand. In practice, most if not all of these lines will have to be chosen by trial and error, guided by any knowledge of the general shape of the underlying surface. However, those associated with the curves of discontinuity, and with the boundary curves, will often be available, or obtainable by an initial stage of curve fitting to discrete data. Sometimes functions specifying the ordinate values along the curves of discontinuity will also be obtainable in the same way.

The least-squares fitting of a bicubic spline with curved knot lines to a given set of data points (X_R, Y_R, Z_R), $R = 1, 2, \ldots, m$, can be carried out in a manner closely similar to that for a conventional bicubic spline, one with constant knots. The latter case was treated in [6] and the main part of the computation consists of computing the values of the B-splines at all the data points so as to form the matrix A of the observation equations

$$A\gamma = Z, \tag{6.1}$$

and then reducing A to upper triangular form by orthogonal transformation, prior to a back-substitution process. With the data suitably ordered, the matrix A, the Rth row of which contains the B-spline values at the Rth data point, takes the block-banded form

$$A = \begin{bmatrix} A_{11} & A_{12} & A_{13} & A_{14} & & & \\ & A_{22} & A_{23} & A_{24} & A_{25} & & \\ & & A_{33} & A_{34} & A_{35} & A_{36} & \\ & & & \cdots & \cdots & \cdots & \\ & & & & \cdots & \cdots & \cdots \end{bmatrix} \tag{6.2}$$

in which each A_{rs} is a rectangular matrix, itself of band form with bandwidth 4.

In that earlier paper, Householder transformations were used for the reduction process, but Givens rotations are now preferred. They allow the computations to be organized so that the (possibly large) matrix A is never stored in its entirety, only an individual row at a time: each row is incorporated into the reduction process immediately it is formed.

The only modification required when using curved knot lines occurs in the evaluation of the B-splines at each data point, to form a row of A. The knot values needed for this purpose, instead of being already available from a fixed array, have first to be computed from the equations of the knot lines. The algorithm is otherwise the same as for a conventional bicubic spline, and the matrix A has the same form. Of course, when fitting a bicubic spline with curved knot lines which is to have a ruled segment, it will be necessary to have an algorithm capable of applying linear equality constraints, in most cases at least. In fact, reference [6] does contain such an algorithm for the conventional bicubic spline, and this carries over to the case of curved knot lines with just the same modification as for the unconstrained algorithm. We note, however, that the constrained algorithm destroys the band form of A and so can be much more time-consuming than the unconstrained algorithm.

The precise mode of application of these modified algorithms when a ruled segment is required will depend on the particular data provided. The situation is more clearly exposed if, in forming the matrix A, we order the terms in (5.2) in the alternative manner to that used in [6]. Specifically, we order the terms so that the values of the product $N_i(x;\lambda(y))N_j(y;\mu)$ are entered into column $j+k(i-1)$. Thus the sth block-column of A consists of k individual columns containing the values of those products which have $i=s$ and $j=1, 2, \ldots, k$. The block-band form in (6.2) is achieved if the data points are ordered so that the rth block-row of A relates to the data points lying in the region between the knot lines $\lambda_{r+1}(y)$ and $\lambda_{r+2}(y)$. (A point lying exactly on a knot line is allocated to one of the non-null regions adjacent to it.)

Now, three coincident knot lines contain two null regions and so, in the ruled segment case, block-rows t-3, t-2, t and t+1 are null. The matrix A in (6.2) then takes the following form (in which each sub-matrix A_{rs} is denoted simply by an X, and only block-rows t-4, t-1 and t+2, now adjacent, are shown explicitly).

$$\begin{bmatrix} \cdots\cdots\cdots & & & & & & & & & & \\ & X & X & X & X & & & & & & \\ & & & & X & X & X & X & & & \\ & & & & & & & X & X & X & X \\ & & & & & & & & & & \cdots\cdots\cdots \end{bmatrix} \qquad (6.3)$$

The four sub-matrices in the central block-row of this diagram occupy block-columns t-1 to t+2. The matrix of the constraint equations (5.3) has the form

$$\begin{bmatrix} -\frac{2}{3}I & I & 0 & -\frac{1}{3}I \\ -\frac{1}{3} & 0 & I & -\frac{2}{3}I \end{bmatrix} \qquad (6.4)$$

in which the entries are again in block-columns t-1 to t+2, and I denotes the $k \times k$ unit matrix.

We now consider three situations, differing in respect of the character of the data provided.

(a) Only data points given, including some between the triple knots.

This case is straightforward. It is simply a matter of employing the constrained algorithm with the given data points and the constraints (5.3).

(b) Only data points given, none between the triple knots.

In this case, the constraint equations can be treated as if they were additional observation equations, and the problem solved more efficiently by means of the unconstrained algorithm, adapted for the purpose. Because of the absence of data between the triple knots, the central block-row in (6.3) disappears and so the matrix, in view of its band structure, separates into two independent parts, linked only by the constraint matrix (6.4). More particularly, block-columns t and $t-1$ of A now contain only zeroes, and so the $2k$ coefficients $\gamma_{t,j}$ and $\gamma_{t+1,j}$, with $j=1, 2, \ldots, k$, no longer occur in the observation equations. On the other hand, these coefficients do occur in the $2k$ constraint equations, and associated with a non-singular matrix, namely a unit matrix, as can be seen in (6.4). Therefore, even if they are treated only as observation equations, the least-squares process will ensure that they are satisfied exactly.

Furthermore, they will not affect the determination of the other coefficients, and so an alternative strategy would be to deal separately with each of the two independent sets of genuine observation equations. Specifically, this means using the unconstrained routine in turn with (i) the data points and knot lines on and to the left of $\lambda_t(y)$, and (ii) the data points and knot lines on and to the right of $\lambda_{t+1}(y)$. All the knots μ_j, $j=-1, 0, \ldots, k+2$, would also be needed in both cases. This determines all the γ_{ij} except those with $i=t$ and $t+1$, which are finally obtained from the constraint equations. Since the number of arithmetic operations in solving an over-determined band system is roughly proportional to the number of rows times the square of the band-width, the two strategies would be of comparable efficiency.

(c) Data points given, together with a cubic spline defining the ordinate values along each of the triple knot lines. The knots of both these splines must be the μ_j of the surface.

Here, in effect, we are given the $\gamma_{t-1,j}$ and the $\gamma_{t+2,j}$, and so, through (5.3), the $\gamma_{t,j}$ and $\gamma_{t+1,j}$ also. This case can be solved in a single application of the constrained algorithm, simply by adding to the computation in (a) above the constraints

$$\begin{aligned} \gamma_{t-1,j} &= \bar{\gamma}_{t-1,j}, \\ \gamma_{t+2,j} &= \bar{\gamma}_{t+2,j}, \qquad j = 1, 2, \ldots, k, \end{aligned} \tag{6.5}$$

where the bars indicate given values. However, the given γ_{ij} completely determine the ruled segment, and so the data points on and between $\lambda_t(y)$ and $\lambda_{t+1}(y)$ have no effect on the solution and can be discarded. Thus the computation can again be split into two parts, and here this alternative would be more efficient. Formally, it is simply a matter of substituting the given γ_{ij} into the observation equations and then proceeding to a least-squares solution for each of the two subsets of modified equations. In terms of general fitting algorithms, the constrained algorithm could be employed with each set of data (i) and (ii), defined in (b) above, together with the first or second set of constraints in (6.5), as appropriate. However, the constraints are now boundary constraints, and an algorithm designed specifically to deal with such constraints would be much more efficient, since band structure is preserved. Such an algorithm would be an extension to two variables of the "method of data modification" described in [2].

The above discussion on the three cases needs modification if, as described at the end of Section 5, the two triple knot lines coincide for part of their range and we wish to preserve continuity of the surface. In that event, we must satisfy the additional constraints (5.4) and as a result the problem never splits into two parts. We can, however, apply the constrained algorithm to the whole data in all cases. The other change is that some of the B-spline products are now zero everywhere, but this can be ignored. The constraints in (5.3) which contain the coefficients of these products are no longer necessary, but it is simpler to leave them in and they do no harm: indeed their inclusion avoids the introduction of spurious rank deficiencies into the linear system.

REFERENCES

[1] Cox, M.G. The numerical evaluation of B-splines. J. Inst. Maths Applics, 10, 134-149, 1972.
[2] Cox, M.G. The incorporation of boundary constraints in spline approximation problems. Lecture Notes in Mathematics 630: Numerical Analysis, ed. by G.A. Watson, Springer-Verlag, Berlin, 51-63, 1978.
[3] Curry, H.B. and Schoenberg, I.J. On Pólya frequency functions IV: the fundamental spline functions and their limits. J. Analyse Math., 17, 71-107, 1966.
[4] De Boor, C. On calculating with B-splines. J. Approximation Theory, 6, 50-62, 1972.
[5] Greville, T.N.E. On the normalization of the B-splines and the location of the nodes for the case of unequally spaced knots. Inequalities, ed. by O. Shisha, Academic Press, New York and London, 286-290, 1967.
[6] Hayes, J.G. and Halliday, J. The least-squares fitting of cubic spline surfaces to general data sets. J. Inst. Maths Applics, 14, 89-103, 1974.
[7] Hayes, J.G. New shapes from bicubic splines. Proceedings CAD 74, Imperial College, London. IPC Business Press, Guildford, fiche 36G/37A, 1974. Also, National Physical Laboratory Report NAC 58, 1974.

ON THE TIME INTEGRATION OF PARABOLIC DIFFERENTIAL EQUATIONS

P.J. van der Houwen

The first part of this paper deals with the time integration of parabolic problems by explicit, stabilized Runge-Kutta methods. These methods are constructed in such a way that arbitrarily large integration steps are allowed without unstable behaviour. The main tool is the application of Chebyshev polynomials. We will relate such Runge-Kutta methods with Richardson's iterative method which leads us to the application of these Runge-Kutta methods for the solution of implicit equations. In the second part of the paper we will use them in solving the fine and coarse grid problems arising in a multigrid algorithm. A few comparative experiments are reported.

1. INTERNAL STABILITY

Let the parabolic initial-boundary value problem be semi-discretized to obtain an initial value problem for the system of ordinary differential equations

(1.1) $$\frac{dy}{dt} = f(t,y).$$

We will assume that the eigenvalues of $\partial f/\partial y$ are situated in a long, narrow strip along the negative axis. For notational simplicity we use the autonomous representation of (1.1) by considering t as the first component of the vector y.

Consider the class of Runge-Kutta methods which can be written in the form

(1.2) $$y_{n+1}^{(0)}=y_n,\quad y_{n+1}^{(j)}=\lambda_j y_n+\tilde{\lambda}_j \Delta t f(y_n)+\mu_j y_{n+1}^{(j-1)}+\tilde{\mu}_j \Delta t f(y_{n+1}^{(j-1)})+\nu_j y_{n+1}^{(j-2)},\quad j=1,2,\ldots,m,$$

where $\nu_1 = 0$. Here, y_n and $y_{n+1}=y_{n+1}^{(m)}$ denote the numerical approximations to $y(t_n)$ and $y(t_{n+1})$; Δt is the integration step $t_{n+1}-t_n$.

The linear stability theory applied to (1.2) leads to a condition on Δt of the form

(1.3) $$\Delta t < \frac{\beta}{\sigma},\quad \sigma \text{ spectral radius of } \partial f/\partial y,$$

where β represents the (real) *stability boundary* of (1.2). The *stability polynomial* $R_m(z)$ of (1.2) generated by the recurrence relation

(1.4) $$R_0(z)=1,\quad R_j(z)=\lambda_j+\tilde{\lambda}_j z+(\mu_j+\tilde{\mu}_j z)R_{j-1}(z)+\nu_j R_{j-2}(z),\quad j=1(1)m$$

satisfies the condition $|R_m(z)|\leq 1$ for $-\beta\leq z\leq 0$. For large values of m(m=100 say), however, the condition (1.3) does not guarantee a stable behaviour of the Runge-Kutta method. Therefore, we would like to require that the recurrence relation (1.2) itself is stable with respect to perturbations in $y_{n+1}^{(j)}$, i.e. we wish that the variational equation

(1.5) $$\Delta y_{n+1}^{(0)}=\delta^{(0)},\quad \Delta y_{n+1}^{(j)}=[\mu_j + \tilde{\mu}_j \Delta t \frac{\partial f}{\partial y}(y_n)]\Delta y_{n+1}^{(j-1)}+\nu_j \Delta y_{n+1}^{(j-2)}+\delta^{(j)},\quad j=1(1)m$$

is stable for large values of m and an arbitrary sequence of rounding errors $\delta^{(j)}$.

If the coefficients $\mu_j, \tilde{\mu}_j$ and ν_j are constant it can be shown (e.g. STETTER [17, p.205]) that Δy_{n+1} is a linear accumulation of the $\delta^{(j)}$ provided that the eigenvalues of the characteristic equation of (1.5) are in the closed unit disk with no multiple roots on the unit circle. In that case we will call (1.2) *internally stable with respect to perturbations in* $y_{n+1}^{(j)}$. If the coefficients in (1.5) are not constant we may impose a necessary condition : $-1 \leq \nu_j \leq 1 - \min\{|\mu_j|, |\mu_j - \beta\tilde{\mu}_j|\}$. Rigorous statements can only be made for the case $\delta^{(j)} = 0$, $j \geq 1$. Then

$$\Delta y_{n+1}^{(j)} = P_j(\Delta t \frac{\partial f}{\partial y}(y_n))\Delta y_{n+1}^{(0)},$$

where the *internal stability polynomials* $P_j(z)$ are defined by

(1.6) $\qquad P_0(z)=1,\ P_j(z)=(\mu_j+\tilde{\mu}_j z)(P_{j-1}(z)+\nu_j P_{j-2}(z),\ j=1(1)m.$

If $|P_j(z)| \leq 1$ for $-\beta \leq z \leq 0$ and $j=0(1)m$, and if the condition on ν_j is satisfied, then (1.2) will be called *internally stable*).

We will restrict our considerations to first and second order accurate methods. It can be shown that (1.2) is first order accurate if $R_m(0) = R_m'(0) = 1$ and second order accurate if in addition $R_m''(0) = 1$. Stability polynomials satisfying these conditions will be called *first* and *second order consistent*, respectively.

2. STABILITY POLYNOMIALS

In view of the large values of σ which arise in cases where (1.1) originates from the semi-discretization of a parabolic problem, large stability boundaries are desired in order to satisfy the stability condition (1.3) without restricting Δt to unrealistic small values. This problem was considered by several authors. First order consistent stability polynomials $R_m(z)$ with *optimal* stability boundary were already mentioned by YUAN'CHZHAO-DIN [19] and SAUL'YEV [15]. These polynomials are the shifted Chebyshev polynomials

(2.1a) $\qquad R_m(z) = T_m(1 + \frac{z}{m^2})$

with stability boundary

(2.1b) $\qquad \beta = 2m^2.$

In actual computation, it is desirable to use strongly stable stability polynomials which are less than 1 in magnitude in the (open) stability interval $(-\beta,0)$. Such polynomials can be derived from (2.1) by writing [7]

(2.2a) $\qquad R_m(z) = \dfrac{T_m(w_0+w_1 z)}{T_m(w_0)}\ ,\ w_0 > 1,$

where w_1 is determined by the condition $R_m'(0)=1$ and w_0 defines the value of the extrema of $R_m(z)$ in the interval $[-\beta,0)$. For $(w_0-1)m^2 \ll 1$ these extreme values (denoted by $\pm$ D) and the stability boundary β are approximately given by

(2.2b) $\qquad \beta = \dfrac{w_0+1}{w_1} \cong 2m^2[1 - \dfrac{4m^2-1}{6}(w_0-1)],\ D = \dfrac{1}{T_m(w_0)} \cong 1-(w_0-1)m^2.$

Examples of optimal stability polynomials of higher order were given by METZGER [13], LOMAX [11] and RIHA [14]. Strongly stable versions for $m \leq 12$ may be found in [7]. The results in these papers indicate that the stability boundary for second order consistent polynomials is given by

$$\beta = c(m)m^2, \quad c(m) \uparrow .82 \quad \text{as } m \to \infty, \tag{2.3}$$

where $c(m)$ is a slowly varying function. The strongly stable versions possess a stability constant $c(m)$ which is slightly smaller.

From the results above it may be concluded that the *effective* or *scaled stability boundary* $\beta^* = \beta/m$ (cf. [7,10]) increases almost linearly with m. In order to exploit this property in actual integration of a parabolic problem, one wishes to use large values of m and therefore explicit expressions for the stability polynomials $R_m(z)$ are required for arbitrary values of m. Unfortunately, such expressions are not available for the optimal polynomials of order 2 or higher. It is possible, however, to construct "almost" optimal stability polynomials of second order in explicit form.

Firstly, consider the polynomials [8]

$$R_m(z) = a+(1-a)\,\frac{T_m(w_0+w_1z)}{T_m(w_0)}, \tag{2.4a}$$

where a and w_1 are determined by the conditions $R_m'(0)=R_m''(0)=1$ and w_0 again defines the damping parameter D of $R_m(z)$. For $w_0=1$ these polynomials were given by BAKKER (cf. [7]). For $w_0 \cong 1$ we find

$$\beta = \frac{w_0+1}{w_1} \cong \frac{2}{3}(m^2-1)[1-\frac{2}{15}(w_0-1)(m^2-\frac{1}{4})], \quad D = a + \frac{1-a}{T_m(w_0)} \cong 1-\frac{1}{3}(w_0-1)(m^2-1). \tag{2.4b}$$

The polynomials (2.4), although not optimal, cover roughly 80% of the stability interval of the optimal stability polynomials (cf. (2.3)).

Next consider polynomials of the form

$$R_m(z) = 1 + \frac{az}{1-bz}\,[T_m(w_0+w_1-w_1bz)-T_m(w_0)]. \tag{2.5}$$

The first order condition $R_m'(0)=1$ expresses the coefficient a in terms of w_0 and w_1, and the second order condition $R_m''(0)=1$ determines the coefficient b i.e.,

$$a = \frac{1}{T_m(w_0+w_1)-T_m(w_0)}, \qquad b = \frac{1}{2[1-w_1aT_m'(w_0+w_1)]}. \tag{2.6}$$

The parameters w_0 and w_1 are free for maximizing the real stability interval. For $a<0$, $b>0$ and $w_1<0$ it can be verified that the interval in which $|R_m(z)| \leq D$ is determined by the inequalities

$$\frac{1+w_0+w_1}{bw_1} \leq z \leq \frac{w_0+w_1-1}{bw_1}$$
$$[(T_m(w_0) \pm 1)a + (1\pm D)b]z \lesseqgtr 1 \pm D \tag{2.7}$$

It is convenient to use instead of the parameters w_0 and w_0+w_1 the variables θ and γ defined by

$$w_0+w_1 = \cos\frac{\theta}{m}, \qquad w_0 = \cosh\frac{\gamma}{m}.$$

The following theorem is now immediate from (2.7).

THEOREM 2.1. *The polynomials* (2.5) *satisfy* $|R_m(z)| \le D$ *in the interval*

$$-\beta = -\frac{1 \cos \theta/m}{b(\cosh\frac{\gamma}{m} - \cos\frac{\theta}{m})} \le z \le \frac{1-D}{a(\cosh\gamma-1)+b(1-D)}$$

provided that $D \ge \pm [1 + \frac{a}{b}(\cosh\gamma \pm 1)]$. □

As an application of this theorem we put $\gamma=0$ and $a=-b$ to obtain the *first order* polynomials ($a=-b=1/(\cos\theta-1)$)

$$R_m(z) = 1 - \frac{z}{1-\cos\theta-z}\left[T_m\left(\cos\frac{\theta}{m} + \frac{\cos\frac{\theta}{m}-1}{\cos\theta-1}z\right)-1\right],$$

(2.8)

$$D = 1, \ \beta = \frac{1-\cos\theta}{\operatorname{tg}^2\frac{\theta}{2m}} \cong 4\,\frac{1-\cos\theta}{\theta^2}\,m^2 \quad \text{as } m \to \infty.$$

This family of polynomials is weakly stable (strongly stable polynomials are obtained if we choose $\gamma > 0$). The cases $\theta=0$ and $\theta=\pi$ are of particular interest. For $\theta=0$ the stability boundary β reaches the maximal value $2m^2$ and the polynomial $R_m(z)$ becomes identical to the optimal polynomial (2.1a). For $\theta=\pi$ the polynomials (2.8) are *second order consistent*. The stability boundary β is then given by

(2.8b') $$\beta = \frac{2}{\operatorname{tg}^2\frac{\pi}{2m}} \uparrow \frac{8}{\pi^2}m^2 \cong .8106m^2 \text{ as } m \to \infty.$$

A comparison with the stability boundary (2.3) experimentally derived for the optimal second order polynomials reveals that the corresponding polynomials (2.8) are practically optimal.

3. INTERNALLY STABLE RUNGE-KUTTA METHODS

A method occasionally used in the literature (e.g. SAUL'YEV [15] and GENTZSCH and SCHLÜTER [4]) is of the "factorized" form

(3.1) $$y_{n+1}^{(0)}=y_n, \ y_{n+1}^{(j)}=y_{n+1}^{(j-1)}+\tilde{\mu}_j\Delta t f(y_{n+1}^{(j-1)}), \quad j = 1,2,\dots,m.$$

This scheme can be adapted to any polynomial $R_m(z)$ by identifying $-\tilde{\mu}_j$ with the reciprocals of the zeros of $R_m(z)$. A disadvantage of (3.1) is the necessity to have real zeros in $R_m(z)$ in order to keep $\tilde{\mu}_j$ real. Thus, (2.2a) is a suitable stability polynomial but (2.4a) is not. This disadvantage can be overcome by defining the "diagonal" scheme [7]

(3.2) $$y_{n+1}^{(0)}=y_n, \ y_{n+1}^{(j)}=y_n+\tilde{\mu}_j\Delta t f(y_{n+1}^{(j-1)}),$$

which can be adapted to any $R_m(z)$ by choosing $\tilde{\mu}_j=\beta_{m-j+1}/\beta_{m-j}$ where β_j denotes the coefficient of z^j in $R_m(z)$.

Although extremely simple, both schemes are unattractive because a substantial part of the internal stability polynomials $P_j(z)$ are highly unstable. Experiments reported in [8] show that these schemes are unreliable for larger values of m(m>12).

It is possible to construct less simple, but internally stable methods according to the following approach. We start with a suitable, prescribed stability polynomial $R_m(z)$ and define generating polynomials $R_j(z)$, $j < m$ with a certain degree of freedom in their coefficients. Next we derive the recurrence relation (1.4) and deduce the internal stability polynomials defined by (1.6). Finally, the freedom in the coefficients of $R_j(z)$ is used for making the method internally stable, either with respect to $y^{(0)}_{n+1}$ or, if the coefficients $\mu_j, \tilde{\mu}_j$ and ν_j are constant, with respect to all $y^{(j)}_{n+1}$. The crucial point in this approach is the possibility to find a recurrence relation of the form (1.4) for the generating polynomials $R_j(z)$. This can be achieved by expressing $R_j(z)$ in terms of orthogonal polynomials such as the Chebyshev polynomials (compare a similar situation in the derivation of stable polynomial iteration methods for elliptic problems [2,3]).

If we succeed in constructing an internally stable method with stability boundary $\beta \cong cm^2$ then there are no (stability) restrictions to the integration step Δt provided that m is sufficiently large, i.e.,

$$m \geq \sqrt{\frac{\sigma\Delta t}{c}}\,. \tag{3.3}$$

In the cases (2.2) and (2.4) the construction of internally stable methods may be found in [8]. Here, we consider the stability polynomial (2.5) and define the generating polynomials

$$R_j(z) = 1 + \frac{a_j z}{1-bz}\,[T_j(w_0+w_1-w_1bz)-T_j(w_0)], \qquad j = 0(1)m \tag{3.4}$$

where a_m=a. These polynomials satisfy the recurrence relation

$$\begin{aligned} &R_0(z)=1,\ R_1(z)=1+a_1w_1z,\\ &R_j(z)=a_j\Big[\frac{1}{a_j} - 2\,\frac{w_0+w_1}{a_{j-1}} + \frac{1}{a_{j-2}}\Big] + 2a_jw_1\Big[\frac{b}{a_{j-1}} + T_{j-1}(w_0)\Big]z\\ &\qquad + 2\frac{a_j}{a_{j-1}}(w_0+w_1-w_1bz)R_{j-1}(z) - \frac{a_j}{a_{j-2}}R_{j-2}(z), \quad j \geq 2. \end{aligned} \tag{3.5}$$

The internal stability polynomials are given by

$$\begin{aligned} &P_0(z)=1,\ P_1(z) = \frac{a_1}{a_0}(w_0+w_1-w_1bz),\\ &P_j(z)= \frac{a_j}{a_0}\,T_j(w_0+w_1-w_1bz), \quad j \leq 2. \end{aligned} \tag{3.6}$$

Following a suggestion of SOMMEIJER [16], we may try to choose $R_j(z)$ such that $y^{(j)}_{n+1}$ is second order accurate in some point $t^{(j)}_{n+1}$ for $j \geq 2$. This is achieved if $R_j(z)$ satisfies the conditions

$$R_j(0)=1, \quad [R'_j(0)]^2 = R''_j(0), \quad j \geq 2$$

which leads to

$$(3.7a)\qquad a_j = 2b\,\frac{T_j(w_0+w_1)-T_j(w_0)-w_1T_j'(w_0+w_1)}{[T_j(w_0)-T_j(w_0+w_1)]^2}\quad,\quad j\geq 2,$$

where b is defined by (2.6) (notice that a_m=a). In order to get internal stability we choose $a_0=a_1=\min a_j$ with j=2(1)m. It is easily verified that (3.7a) yields almost constant coefficients $(\mu_j,\tilde{\mu}_j,\nu_j) = (2w_0+2w_1,-2w_1b,-1) + O(\frac{1}{m})$.

A second possibility which is attractive from a computational point of view, consists in choosing the coefficients a_j such that the coefficients $\mu_j,\tilde{\mu}_j$ and ν_j are constant. This is achieved by putting

$$(3.7b)\qquad a_j = aq^jT_m(w_0),\quad q = \frac{1}{[T_m(w_0)]^{1/m}},\ j = 0(1)m$$

which results in

$$\lambda_1=1,\ \tilde{\lambda}_1 = aw_1qT_m(w_0),\ \mu_1=\tilde{\mu}_1=\nu_1 = 0,$$

$$(3.8)\qquad \lambda_j=1-2(w_0+w_1)q+q^2,\ \tilde{\lambda}_j=2w_1q[b+aq^{j-1}T_{j-1}(w_0)\,T_m(w_0)],$$

$$\mu_j=2(w_0+w_1)q,\ \tilde{\mu}_j=-2w_1qb,\ \nu_j=-q^2,\ j=1(1)m.$$

We recall that the corresponding scheme is first order if a is given by (2.6) and second order accurate if b is also given by (2.6). Since $q\leq 1$ for $w_0\geq 1$, the internal stability polynomials $P_j(z)$ are stable with damping factor q^j for $-\beta\leq z\leq 0$. We also have stability with respect to perturbations in all stages because the characteristic equation corresponding to (1.5), i.e. the equation

$$(3.9)\qquad \zeta^2-2q(w_0+w_1-w_1bz)\,\zeta+q^2=0,\qquad -\beta\leq z\leq 0,$$

has its roots within the unit circle for $w_0>1$ and $w_0+w_1<1$.

4. RICHARDSON'S METHOD

Let us choose in (3.4)

$$(4.1)\qquad a_j=a\,\frac{T_m(w_0)}{T_j(w_0)},\ j = 0(1)m,$$

to obtain the internal damping factor $T_j^{-1}(w_0)$. The scheme can be written in the form

$$y_{n+1}^{(0)} = y_n,$$

$$(4.2)\qquad y_{n+1}^{(j)} = \mu_jy_{n+1}^{(j-1)}+(1-\mu_j)y_{n+1}^{(j-2)}+\mu_j^*[\textstyle\sum - y_{n+1}^{(j-1)}+b\Delta tf(y_{n+1}^{(j-1)})],\quad j = 1(1)m$$

$$\textstyle\sum\quad = y_n-[b+aT_m(w_0)]\Delta tf(y_n),\quad \mu_1=1,$$

where $\mu_j=2w_0T_{j-1}(w_0)/T_j(w_0)$ and $\mu_j^*=-w_1\mu_j/w_0$. As before, the scheme (4.2) is first or second order accurate if a or a and b are defined by (2.6). However, other choices are possible by interpreting (4.2) as an *iteration method* for solving the equation

(4.3) $\qquad y-b\Delta t f(y) = \sum$.

The iteration error $y_{n+1}^{(j)} - \eta$, η being the solution of (4.3), is approximately given by [9]

(4.4) $$y_{n+1}^{(j)} - \eta = P_j(\Delta t \frac{\partial f}{\partial y}(y_n))(y_n-\eta).$$

Suppose that w_0 and w_1 are chosen such that $P_j(z)$ has a minimal maximum norm in the eigenvalue interval of $\Delta t\partial f/\partial y$, then the resulting method (4.2) is easily recognized as *Richardson's iteration method* applied to (4.3) with y_n as initial approximation. This raises the question: why not starting from a first or second order implicit formula of the form (4.3), approximating its solution by m Richardson iterations and considering $y_{n+1}=y_{n+1}^{(m)}$ as the numerical approximation at t_{n+1}. We will show that the resulting m-stage Runge-Kutta method may indeed be of interest.

Let in (4.2)

(4.5) $$a = -T_m^{-1}(w_0), \quad b \geq \tfrac{1}{2},$$

then (4.3) is A-stable and first order consistent (if $b=\frac{1}{2}$ we have second order consistency). The parameters w_0 and w_1 are defined by

(4.6) $$w_0+w_1 = \frac{\sigma+\delta}{\sigma-\delta}, \quad w_1=\frac{-2}{b(\sigma-\delta)\Delta t} .$$

Here, δ and σ are the smallest and largest eigenvalue in magnitude of $\partial f/\partial y$ (if f was made autonomous by introducing t as an additional component of the vector y we put $t_{n+1}^{(j)} = t_{n+1}$ throughout the iteration process, otherwise we would have an eigenvalue zero in the spectrum of $\partial f/\partial y$ and therefore $\delta=0$). Applying theorem 2.1 we conclude that the resulting method is stable if

(4.7) $$\cosh \gamma \geq \frac{1}{2b-1} \text{ where } \cosh \frac{\gamma}{m} = \frac{\sigma+\delta+\frac{2}{b\Delta t}}{\sigma-\delta} .$$

This stability condition yields a lower bound for m (cf. 3.3)). Let $\delta<<\sigma$ and let us suppose that after m iterations we want a reduction of the iteration error by a factor $\rho = T_m^{-1}(w_0)=1/\cosh \gamma$. Then the stability condition (4.7) yields

(4.7') $$m \cong \sqrt{\frac{\Delta t\sigma}{c}}, \quad c \cong 4 \frac{1+b\delta\Delta t}{b[\operatorname{arccosh} \frac{1}{\rho}]^2}, \quad b \geq \tfrac{1}{2}(1+\rho).$$

Thus, if we choose the trapezoidal rule as our starting formula ($b=\frac{1}{2}$) the scheme (4.2) is unstable unless $\rho=0$ which leads to extremely large values of m. For $b>\frac{1}{2}$, e.g. b=1 (backward Euler), and a reduction factor $\rho=1/100$ (say) we obtain a stability "constant" $c \cong (1+\delta\Delta t)/7$. Small values of $\delta\Delta t$ leads to small stability constants, but it is conceivable that in an actual application $\delta\Delta t$ is so large that c is even larger then the optimal stability constant of first order Runge-Kutta methods where c=2. The reason is of course that by (4.5) the implicit method (4.3) becomes first order consistent, *not the numerical scheme (4.2)*. Nevertheless, if high accuracies are not necessary, Richardson's method might be of interest in special problems.

A disadvantage is the requirement to provide both σ and δ in order to get an optimal reduction of the iteration error (the stabilized Runge-Kutta methods discussed in the preceding sections only require σ). A second unfavourable property of Richardson's method turns out to be the large error constants if time-dependent boundary conditions are introduced [9]. This indicates that Richardson's method is less suited as an integration method on its own. In the next section, however, we will see that it may be used as part of a multigrid method where the possibility to adapt the polynomial $P_j(z)$ to the equation to be solved can be fully exploited.

5. A TWO-GRID METHOD

In the preceding sections *explicit* time integration of parabolic problems was discussed. Such methods are rather robust, in particular for highly nonlinear problems, but inspite of the large stability boundaries, they still require relatively many right hand side evaluations per integration step. In this section we will construct a two-grid method in which the *implicit relations* are solved by the *explicit methods* discussed before.

We start with the implicit method of first order (cf. (4.3))

$$(5.1)\qquad y-b\Delta t\ f(y) = y_n+(1-b)\Delta t f(y_n) =: \textstyle\sum_h,$$

where b is a free parameter (b=1 yields implicit Euler, b=½ the trapezoidal rule). Let (5.1) be defined on a two-dimensional spatial grid Ω_H then we define on a coarser grid Ω_H the coarse grid problem [1,5]

$$(5.2)\qquad y-b\Delta t\ f_H(y) = R \textstyle\sum_h + b\Delta t[Rf(\hat{y}_h) - f_H(R\hat{y}_h)] =: \textstyle\sum_H.$$

where f_H denotes the right hand side function on the grid Ω_H, R is the *restrictor* which restricts a grid function defined on Ω_h to a grid function defined on Ω_H (for a definition see [6]), and $\hat{y}_h$ denotes an approximation to the solution of (5.1). The coarse grid problem (5.2) may be considered as the restriction of a slightly perturbed fine grid problem.

The fine grid problem (5.1) is iteratively solved [5] by performing alternatingly iterations on the coarse grid and on the fine grid. Let $\hat{y}_h$ denote the last approximation to the solution of (5.1), then the coarse grid iteration is started with $R\hat{y}_h$, and if y_H^* is the solution of (5.2), then the fine grid iteration is started with

$$(5.3)\qquad \hat{y}_h + P(y_H^* - R\hat{y}_h),$$

where P is the prolongator which prolongates a grid function defined on Ω_H to a grid function defined on Ω_h (we used linear interpolation [6]). The second term in (5.3) is called the *coarse grid correction* and can be interpreted as a defect correction (cf. [18,5].

Here, we restrict our considerations to *two-grid methods*. It is of course possible to apply the method just described recursively to obtain a full multigrid method.

For a detailed discussion of such multigrid methods we refer to [1,5,6].

The idea of a two-grid method is the *reduction* of the *short wave lengths* in the iteration error by iteration on the *fine grid* and the *long wave lengths* by iteration on the *coarse grid*. In general, an iteration method has no difficulties in reducing the short waves, but has a relatively poor damping effect on the long waves. The use of coarse grid iteration, however, enables us to reduce waves which are long on the fine grid but shorter with respect to the coarse grid. Thus, we need an iterative method in which the damping of the waves can be tuned to the coarse and the fine grid problem. Such a method is Richardson's method where the damping of a particular part of the system of eigenvectors can be controlled by the parameters w_0 and w_1 or δ and σ (cf. (4.6)). Of course, Richardson's method can only be applied if the eigenvalues of $\partial f/\partial y$ are more or less negative. For a discussion of Richardson's method for problems where the eigenvalues are located in an ellipse with its main axis along the negative axis, we refer to MANTEUFFEL [12].

In our experiments we used the (nonlinear) Richardson iteration in the form of scheme (4.2). Starting on the *coarse grid*, we chose

$$(5.4)\qquad H=2h,\ \sum = \sum_H,\ b=1,\ \delta= \frac{\sigma}{100}\ ,\quad \rho = \frac{1}{10}\ ,\quad \sigma = \text{spectral radius}\ \frac{\partial f}{\partial y}\ .$$

The parameters w_0, w_1 and m follow from (4.6) and (4.7'). For m we find

$$(5.5)\qquad m \cong \sqrt{\frac{\sigma\Delta t}{.44[1+\frac{\sigma\Delta t}{100}]}} \simeq 15 \quad \text{as } \sigma\Delta t >> 100.$$

The reduction on the coarse grid requires the evaluation of Σ_H plus 15 evaluations of f_H which is roughly equivalent to 5 evaluations of f(y) on the fine grid. In this reduction step, on 99% of the eigenvalue interval the waves are reduced by a factor 1/10 (for a discussion of the choice of these figures we refer to [9]).

On the *fine grid* we chose

$$(5.6)\qquad \sum = \sum_h,\ b=1,\ \delta = \frac{\sigma}{4}\ ,\ \rho = .22,\quad m = 2,$$

which requires 2 evaluations of f in each fine grid reduction.

In order to start this two-grid method one may use Ry_n as initial guess on the coarse grid. However, in the case of time-dependent boundary values, it is recommended to start with a more accurate initial approximation in order to avoid the large error constants introduced by inconsistencies between internal and boundary grid point values of $y_{n+1}^{(j)}$ [9]. In our experiments, we therefore replaced the *first* coarse grid reduction by an application of the scheme (4.2) with

$$(5.4')\qquad H = 2h,\ \sum = \sum_H,\ b = 1,\ w_0+w_1 = \cos\frac{\pi}{2m}\ ,\ w_0 \cong \cosh\frac{1}{3m}\ .$$

This may be interpreted as a first order consistent Runge-Kutta method for solving the coarse grid problem. The value of m was defined by (3.3) where $c = \beta/m^2$ follows from theorem 2.1. A simple calculation yields $m \cong \sqrt{\sigma\Delta t/1.55}$. Generally, this method is more expensive than (5.4) (cf. (5.5)), but the increased accuracy makes (5.4') more efficient.

6. NUMERICAL EXPERIMENTS

Besides the two-grid Richardson method (*TGR method*) just described, we tested several other integration methods. Firstly, the *TGG method*, similar to the TGR method but the Richardson iteration replaced by symmetric Gauss-Seidel iteration. The number of iterations were chosen such that both methods require the same number of right hand side evaluations per integration step, and the (scalar) implicit relations were solved by just one Newton iteration. Secondly, the *Runge-Kutta-Chebyshev (RKC1) method* based on (2.4a) and described in [8,16], and finally, the Runge-Kutta-Chebyshev methods defined by (3.5), (3.7a) and (2.6), (3.5), (3.7b), respectively. These methods will be denoted by *RKC2* and *RKC3*. In the RKC methods the number of stages was chosen according to (3.3) and the internal damping factor $\|P_m(z)\|_\infty$ was set to .95.

As stability test problem we chose the highly nonlinear equation

$$(6.1) \qquad \frac{\partial U}{\partial t} = \Delta(U)^5, \qquad 0 \le t \le 1, \quad 0 \le x_1, x_2 \le 1$$

with Dirichlet boundary conditions and initial condition to be derived from the exact solution

$$(6.2) \qquad U(t,\vec{x}) = [\tfrac{4}{5}\,(2t+x_1+x_2)]^{1/4}.$$

By the usual 5-point discretization of Δ on a uniform grid with mesh width h, this problem was discretized to obtain a system of the form (1.1). The eigenvalues of $\partial f/\partial y$ are approximately located in the interval

$$(6.3) \qquad [-\frac{64(1+t)}{h^2}, \; -16\pi^2(t+h)].$$

Evidently, this problem is highly stiff ($h=1/20$ yields $\sigma \cong 25600(1+t)$) while the solution is rather smooth. Hence, the integration steps will hardly be restricted by accuracy and therefore (6.1) is a suitable test problem for stability.

For the TGR and TGG methods we listed in table 6.1 the total number N of right hand side evaluations on the grids Ω_h and Ω_{2h} expressed in terms of evaluations on the fine grid Ω_h. Furthermore, we give the accuracy obtained by means of the number of correct digits at t=1, defined by

$$(6.4) \qquad A = \min_{\Omega_h}[-\log_{10}|\text{exact solution (6.2) - numerical solution}|] .$$

Table 6.1 Results obtained for h=1/20, H=1/10

Δt	ν	TGR N	TGR A	TGG N	TGG A
1	1	26	1.7	*	
	2	33	2.1	*	
	3	40	2.6	*	
	4	47	3.0	*	
$\frac{1}{5}$	1	62	2.8	34	1.7
	2	97	4.0	68	2.4
	3	132	4.1	101	2.9
	4	167	4.1	135	3.5

Table 6.2 Results obtained for h=1/20

Δt	RKC1 N	RKC1 A	RKC2 N	RKC2 A	RKC3 N	RKC3 A
1	280	3.0	255	3.2	255	2.7
$\frac{1}{2}$	400	3.3	337	3.5	337	3.2
$\frac{1}{5}$	526	4.1	510	4.2	510	3.6
$\frac{1}{10}$	756	4.9	711	5.5	711	4.1

Instability is denoted by * and the number of coarse grid corrections by ν. The TGG method was unstable for $\Delta t > 1/5$.

Results obtained by the RKC methods are listed in table 6.2.

ACKNOWLEDGEMENT. The author is indebted to Mr. B.P. Sommeijer for carefully reading the manuscript and for carrying out the numerical experiments.

REFERENCES

1. BRANDT, A., *Multi-level adaptive techniques (MLAT) for singular perturbation problems.* In Numerical Analysis of singular perturbation problems, P.W. Hemker and J.J.H. Miller eds., Academic Press, 1979.

2. FRANK, W., *Solution of linear systems by Richardson's method,* J. Assoc. Comput. Mach. 7, 274-286 (1960).

3. FORSYTHE, G.E. & WASOW, W.R., *Finite difference methods for partial differential equations,* John Wiley, New York (1960).

4. GENTZSCH, W. & SCHLÜTER, A., *Über ein Einschrittverfahren mit zyklischer Schrittweitenänderung zur Lösung parabolischer Differentialgleichungen,* ZAMM 58, T415-T416 (1978).

5. HEMKER, P.W., *Introduction to multigrid methods,* Nieuw Arch. voor Wiskunde 29, 71-101 (1981).

6. – , *On the structure of an adaptive multi-level algorithm,* BIT 20, 289-301 (1980).

7. HOUWEN, P.J. VAN DER, *Construction of integration formulas for initial value problems,* North-Holland (1977).

8. HOUWEN, P.J. VAN DER & SOMMEIJER, B.P., *On the internal stability of explicit, m-stage Runge-Kutta methods for large m-values,* ZAMM 60, 479-485 (1980).

9. HOUWEN, P.J. VAN DER & SOMMEIJER, B.P., *Analysis of Richardson iteration in multigrid methods for nonlinear parabolic differential equations*, Report NW105/81, Math. Centrum, Amsterdam 1981 (prepublication).

10. JELTSCH, R. & NEVANLINNA, O., *Stability of explicit time discretizations for solving initial value problems*, Report no. 30, University of Oulu 1979 (prepublication).

11. LOMAX, H., *On the construction of highly stable, explicit, numerical methods for integrating coupled ordinary differential equations with parasitic eigenvalues*, NASA Technical Note NASATN D/4547 (1968).

12. MANTEUFFEL, T.A., *The Tchebyshev iteration for non-symmetric linear systems*, Num. Math. 28, 307-327 (1977).

13. METZGER, CL. *Méthodes de Runge-Kutta de rang supérieur à l'ordre*. These (troisieme cycle), Université de Grenoble (1967).

14. RIHA, W., Optimal stability polynomials, Computing 9, 37-43 (1972).

15. SAUL'YEV, V.K., *Integration of equations of parabolic type by the methods of nets*, Pergamon Press, New York (1964).

16. SOMMEIJER, B.P. & VERWER, J.G., *A performance evaluation of a class of Runge-Kutta-Chebyshev methods for solving semi-discrete parabolic differential equations*, Report NW 91/80, Math. Centrum, Amsterdam (1980).

17. STETTER, H.J., *Analysis of discretization methods for ordinary differential equations*, Springer-Verlag, Berlin (1973).

18. ------., *The defect correction principle and discretization methods*, Num. Math. 29, 425-443 (1978).

19. YUAN'CHZHAO-DIN, *Some difference schemes for the solution of the first boundary value problem for linear differential equations with partial derivatives*, Thesis, Moscow State University (1958).

PRECISION CONTROL, EXCEPTION HANDLING AND A CHOICE OF NUMERICAL ALGORITHMS

T.E. Hull

Abstract

We require more appropriate programming language facilities if we are to implement the numerical processes we would like to have. The needs for precision control and exception handling are particularly urgent, and some specific proposals for meeting these needs are outlined. A number of examples are then used to illustrate the effectiveness of these new facilities, and, in particular, to show what a wide choice of possible numerical processes is made available to the user with the help of these facilities. Finally, brief mention is made of the use of preprocessors, of the design of a new language for numerical computation, and of an arithmetic unit that is now under construction. Experience with these systems helps support the claim that the proposed facilities can be provided both economically and efficiently.

1. Introduction

This talk is motivated by a belief that the numerical processes we would like to implement require more appropriate programming language facilities than we now have, especially for precision control and exception handling. And, moreover, the needed facilities can be provided in a reasonably convenient and economical way.

We begin in section 2 with a brief indication of what kinds of needs there are. Three are identified, two that involve precision control and one that involves exception handling.

Then in section 3 we outline language facilities that are intended to provide for these needs. The specifications for floating-point arithmetic are discussed first. The emphasis is on having arithmetic which is clean and easy to describe, and which also provides for all eventualities (such as overflow and underflow) in a simple and useful way. Then the precision specifications are discussed, a key point being that the precision of variables is prescribed separately from the precision of the operations performed on the variables. Precision can also be changed dynamically. Exception handling is provided for in such a way that the user has complete control over what course of action is to be taken in case an exception does arise within a particular block.

Section 4 is devoted to examples. The main purpose is to show how the language facilities described in section 3 enable the user to implement a very wide variety of numerical processes, providing a choice which may be sufficiently wide for any processes that we might want to implement in practice. At the same time, the language facilities may be seen to be relatively simple.

Some further practical experience with the ideas described in sections 3 and 4 are mentioned briefly in section 5. Two preprocessors have been developed to implement the main features of the precision control facilities. A prototype language incorporating almost all of the language features has been designed, and also tested on further examples. An arithmetic unit to provide complete support for the language has been designed and is currently being constructed.

Our results are summarized in the final section, and we reach a tentative conclusion that the advantages claimed for the proposed language facilities can be provided in a reasonably cost-effective way.

2. Kinds of needs

We begin by identifying the following three needs which arise in various numerical processes we would like to implement.

(1) We may sometimes wish to carry out different parts of a calculation in different precisions. The obvious example here is in the double precision accumulation of a dot product. Other examples include computing a residual vector and computing the Euclidean norm of a vector. (In the latter it may be of as much interest to carry out intermediate stages of the calculation using a wider exponent range as it is to use higher precision.)

(2) On other occasions we may wish to carry out a part of a calculation in more than one precision. One example would be when we want a measure of the effect of roundoff on a part of a calculation: we simply do that part of the calculation in two different precisions and compare the results. Another more general example is one in which we repeat part of a calculation in higher and higher precision until some a posteriori estimate of the error (which is obtained at the end of each repetition) is small enough, at which point the precision does not have to be increased any further.

(3) We may wish to determine what course of action is to be followed in case an exception arises, such as overflow or underflow. For example, in case of overflow we may wish to substitute a large number and carry on with the calculation. Alternatively, we may prefer to stop the calculation after output of an appropriate message. Another possibility is that we may wish to widen the exponent range and do the calculation over again.

3. Language facilities

We assume that floating-point values are normalized, sign and magnitude (for both exponent range and significand), and decimal. The precision is the number of decimal digits in the significand. If the precision is p, we assume the exponent range is [-10p, +10p] - this last assumption is somewhat arbitrary, but it has been found convenient in practice. We also assume that there is a special value called "indeterminate", and another called "not-yet-assigned".

We assume that arithmetic operations are performed "perfectly", in the sense that they produce the properly rounded results, ties being broken by rounding to the nearest even. The wraparound result is produced on overflow or underflow, along with the appropriate interrupt. "Indeterminate" is produced on "zero-divide".

Complex, interval, and complex interval arithmetic would be required in a complete system, but we will restrict our attention almost entirely to "real" arithmetic.

As for the precision of the variables, we assume that they can be declared in an obvious way as illustrated in Figure 1. On the other hand, the precision of the operations performed on the variables is determined by the declared precision of the block in which those operations appear. Thus, in the first block of Figure 1, all operations are carried out in precision 16. The value of x would, in effect, be padded out with 6 zeros. Similarly, the value of y would be padded out and then the sine would be obtained to 16 decimal places, and the sum x + sin(y) would be evaluated to 16 decimal places. The value of u would be rounded to 16 places before the final sum is carried out to 16 places. The result would have to be rounded to 10 places before being assigned to z.

```
real(10) x, y, z                declarations of variables of
real(20) u                         specified precisions
integer p
----
begin precision(16)             precision of all operations in
   real(16) t                      block (except assignment)
   ----                            is to be 16
   z = x + sin(y) + u
   call solve(z,...)
   ----
end
----
begin precision(p)              precision in block is to be p
   real(p)...
   ----
   ----
end
----
begin
   on(overflow)                 overflow handler for block (the
        ----                       block is the scope of the
     (underflow)                   handler)
        result = 0              "result" is a reserved word
   end
   ----                         other exceptions include
   ----                            "inexact" and "error"
end
```

Figure 1. Illustration of language facilities. Some special functions would also be needed, including precisionof( ), currentprecision, lowvalue(,) and highvalue(,). (For example, highvalue(1, A) would be the high value of the first subscript of A.)

Special functions to determine the precision of specific variables, or of the environment, will be needed, as is illustrated in later examples. Special functions will also be needed to determine the lowvalues and highvalues of indices of arrays.

Exception handlers can be prescribed by the user, as is illustrated in the final block in Figure 1. Notice how "result" is a reserved word whose value can be used after an exception has been trapped (for example, it might be the wraparound value after overflow), or whose value can be assigned before the calculation proceeds (for example, it might be assigned the value 0 after underflow, as illustrated in Figure 1).

Earlier indications of these ideas about language facilities, along with some further details, can be found in [4 -7].

4. Examples

We now consider a few examples to illustrate how the language facilities of the preceding section can be used.

The first example (see Figure 2) is a relatively straightforward modification of the LINPACK subroutine SGEFA [3, pp. C.5–C.6] for factoring a single precision

```
procedure GEFA(A, pivot, ind)
   real(*) A(*,*), temp
   integer pivot(*), ind, n, i, j, k, ℓ
   n = highvalue(1,A)
   ind = 0
   for k = 1 to n-1
      find ℓ such that A(ℓ,k) is the pivot
      pivot(k) = ℓ
      if(A(ℓ,k) = 0)
         ind = k
      else
         interchange A(ℓ,k), A(k,k) if ℓ ≠ k
         temp = -1/A(k,k)
         for i = k + 1 to n
            A(i,k) = temp*A(i,k)
         end for
         for j = k + 1 to n
            temp = A(ℓ,j)
            interchange A(ℓ,k), A(k,j) if ℓ ≠ k
            for i = k + 1 to n
               A(i,j) = A(i,j) + temp*A(i,k)
            end for
         end for
      end if
   end for
   pivot(n) = n
   if(A(n,n) = 0)
      ind = n
   end if
end procedure
```

Figure 2. A procedure for factoring the general matrix A, or returning ind ≠ 0 if one of the pivots is exactly 0 (otherwise no provision is made for underflow or overflow). It is intended that the procedure operate at the precision of the environment. (Highvalue(1,A) is the highest value of the first subscript of A.)

general matrix. The modifications are relatively minor ones. For example, we assume that dimensions of arrays do not have to be included in the parameter list. The main point about this example is that, with the language facilities described in the preceding section, the procedure will be executed in whatever precision is specified for the block from which the procedure is called. The name GEFA has been used to help underline the fact that only one version is needed, and it would not be necessary to have a separate version available for each separate precision. (In fact, with complex arithmetic available, it would not be necessary to have separate complex versions available either. Only one, GEFA, would be sufficient for all arithmetics and all precisions.)

Similar modifications could be made to SGESL [3, pp. C.7–C.8] to produce a procedure, say GESL, for solving an already factored system. Figure 3 shows how a procedure for solving linear equations can be constructed with the help of GEFA and GESL. Since no provision has been made for handling overflow or underflow in this procedure, except for the possibility of one of the pivots being exactly 0, any occurrence of overflow or underflow would be referred to the exception handler in force at the time this procedure is called. Otherwise this procedure returns with ind = 0 and the approximate solution x (or with ind not equal to 0 to indicate that one of the pivots is exactly 0).

So far the only significant way in which we have taken advantage of the new language facilities is to show that we need to write only one library procedure for a particular task, regardless of the number of precisions (or even of the different kinds of arithmetic) that one might want the procedure to use. In the next example we will see the new facilities being used in a much more significant way.

In Figure 4 another procedure for solving linear equations is presented. However,

```
procedure linearequations#1(A, b, x, ind)
   real(*) A(*,*), b(*), x(*)
   integer ind, n
   n = highvalue(1,A)
   begin
      integer pivot(n)
      call GEFA(A, pivot, ind)
      if(ind = 0)
         call GESL(A, pivot, b, x)
      end if
   end
end procedure
```

Figure 3. A procedure for finding the (approximate) solution x of Ax = b, or returning a value of ind ≠ 0 if a pivot that is exactly 0 arises in the course of factoring A. (Otherwise no provision is made for handling overflow or underflow.) The procedure assumes the precision of the block from which it is called.

```
procedure linearequations#2(A, b, tol, maxp, x, ind)
   real(*) A(*,*), b(*), tol, x(*)
   integer maxp, ind, p, n
   p = currentprecision
   n = highvalue(1,A)
   ind = -1
   while(ind ≠ 0 & p ≤ maxp)
      begin precision(p)
         real(p) C(n,n), d(n), y(n), rcond
         integer pivot(n)
         on (overflow or underflow)
      ......exit begin block
         end on
         assign arrays C = A, d = b
         call GECO(C, pivot, rcond)
         if(rcond ≠ 0)
            call GESL(C, pivot, d, y)
            if(‖y‖ ≤ 2*10**(-p+1)*g(n)*rcond*tol)
               x = y
               ind = 0
            end if
         end if
      end
      if(p = maxp)
         p = maxp + 1
      else
         p = min(p + 6, maxp)
      end if
   end while
end procedure
```

Figure 4. A second procedure for solving Ax = b. Here, if ind = 0, x will almost certainly be within tol of the true solution. The procedure uses higher and higher precision until the error criterion is satisfied <u>or</u> the user-supplied maximum precision, maxp, would have to be exceeded. (The factor g(n) is a known growth coefficient referred to in [3, p. 1.20].)

the specifications for the procedure are quite different from those in Figure 3. This time the user provides a tolerance, and the procedure attempts to provide an approximate solution that differs in norm from the true solution by no more than the tolerance.

In attempting to accomplish this objective the procedure in Fig.4 uses a modification of LINPACK's SGECO [3, pp. C.1-C.4], called GECO, which produces an estimate "rcond" of the reciprocal of the condition number, as well as a factorization of A. If this reciprocal is ≠ 0, the procedure solves the system and, using the value of rcond, performs a test to determine whether or not the solution it has obtained is sufficiently accurate [3, p. 1.20]. If it is not, the entire process is repeated in higher precision. The repetition can be continued in higher and higher precision until a satisfactory solution is obtained <u>or</u> until some user-supplied maximum precision, maxp, would have to be exceeded.

It must be kept in mind in this example that overflow or underflow could occur when y is assigned to x. However, if the procedure returns with ind = 0, we can be sure that x is within at most a rounding error of a solution satisfying the tolerance requirement, provided only that we can depend on rcond being a sufficiently good estimate.

It would be possible to go one step further and eliminate even this last proviso about rcond. That is, if ind = 0, we could be absolutely certain that the value of x was as close to the true solution as is stated above. However, this would involve the cost of computing an approximate inverse. It would also involve some interval arithmetic, or something essentially equivalent, since guaranteed bounds on the norms of certain residuals are needed and these are most easily calculated with the help of interval arithmetic.

The last example makes much more use of the new language facilities, including both precision control and exception handling facilities. However, the main point to be made in the examples considered so far is that the new facilities provide the user with a wide choice of possibilities. The user can implement any process he wishes, from a straightforward attempt to solve his problem in a conventional way, through an intermediate approach which, with some additional computation and the occasional use of higher precision when necessary, provides him with some assurance about the accuracy of his results, to a more conservative approach which, at additional cost, can provide guaranteed results.

Our examples have all been chosen from linear algebra. But analogous situations occur in other areas. Perhaps the closest area in this respect is zeros of polynomials. Quadrature introduces other factors, but at least it should now be possible to envisage an adaptive quadrature procedure which not only handles subdivisions of the interval of integration in some standard way, but which is also prepared to increase the precision if necessary.

To conclude this section we return to linear algebra and mention just two further examples. One example has to do with determining the least squares solution of an overdetermined system Ax = b. One approach to this problem is to produce the normal equations $A^TAx = A^Tb$, and then to solve this (usually smaller) symmetric system for x. Although mathematically simpler, this approach is usually avoided, because the condition of the normal equations is likely to be much worse than the condition of the original problem. However, this difficulty is easily avoided if the normal equations are formed and solved in double precision. This approach will certainly be simpler and, if double precision is reasonably efficient, it will also be faster than more usual approaches, such as one based on the modified Gram-Schmidt method.

We finish up with just one more example, which is shown in Figure 5. Here the residual r = b-Ax is calculated in a higher precision than the precision used to calculate the approximation to x in Ax = b. The higher precision has been arbitrarily chosen to be 6 digits more than the precision of the environment; it could have as easily been chosen to be twice the precision of the environment. Such residuals are used in various situations. One is in the computation of guaranteed error bounds,

```
procedure residual(A, x, b, r)
   real(*) A(*,*), x(*), b(*), r(*)
   integer n, p
   n = highvalue(1,A)
   p = currentprecision
   begin precision(p + 6)
      real(p + 6) temp
      integer i, j
      for i = 1 to n
         temp = b(i)
         for j = 1 to n
            temp = temp - A(i,j)*x(j)
         end for
         r(i) = temp
      end for
   end begin
end procedure
```

Figure 5. A procedure for calculating the residual vector r = b-Ax in higher precision. The "higher precision" has been arbitrarily chosen to be 6 digits more than the precision of the environment.

as was mentioned earlier. Another is for iterative improvement.

Once again the point is that the user has the facilities available to implement whatever particular process he has in mind, and to do so in a reasonably convenient way.

5. Some practical experience

What has been described so far is intended to indicate what we consider to be "ideal", and a complete implementation of the proposed facilities does not exist as yet. However, we have had some experience with various aspects of the proposed facilities in three different projects which will now be described very briefly.

(1) Two preprocessors have been used to implement some of the basic ideas about precision control. One preprocessor is for a language we now call Algol-H, which is an extension of Algol-W that provides precision control in terms of word lengths (rather than decimal digits), but does nothing about exception handling [6, 7]. The programs produced by this preprocessor are very slow, since every arithmetic operation is translated into a procedure call. Nevertheless, a large number of examples have been tested successfully with this system. A typical result is one where a single precision approximation to the Hilbert matrix of order 8 is used to define a problem for a linear equation solver. The solver had to reach precision 3 before it was able to satisfy an error tolerance of 10**(-8).

The other preprocessor is for a language called Fortran-X. It is similar to the preprocessor just mentioned, except that it is for an extension of Fortran rather than Algol-W, and it produces much more efficient programs since machine arithmetic is used for both single and double precision. The main ideas for this preprocessor were first proposed in [7], but the system itself has been completed only recently.

(2) A complete language called PNCL (for Prototype Numerical Computation Language) has recently been defined [2]. It is based on Pascal and provides almost all of the features proposed in this paper, except that the arithmetic is based on bytes rather than decimal digits. The compiler itself has not yet been written, but the language design has been carefully tested on a number of examples. As a result, some modifications to the language will likely be made before a compiler is developed, but there is good evidence that a reasonably good design is now nearing completion.

(3) An arithmetic unit called CADAC (Clean Arithmetic With Decimal Base And Controlled Precision) has been designed and is currently being constructed. More details are given in [1]. This unit will be able to support all the language facilities proposed in this paper (with the minor exception that the precision must be an even number of decimal digits), and it should do so with reasonable efficiency. (Although somewhat more expensive than conventional arithmetic units, it will handle arbitrarily high precision and, for example, will perform an 8-digit multiply in 4 μs, running at 10 mHz.) Initially, CADAC will be attached to a PDP 11/34. The primary purpose of this unit is to demonstrate the feasibility of supporting the language facilities proposed in this paper in a reasonably efficient and economical way.

6. Summary and conclusions

We began by identifying three needs which are not currently met by most programming languages but which should be provided for if we are not to be restricted in the numerical processes we are able to implement. One need is to carry out part of a calculation in higher precision. Another is to be able to repeat a part of a calculation in higher and higher precision. And the third is to let the user have control over what happens when an exception occurs.

Language facilities to meet these needs were proposed. A key feature of these facilities is the separation of the precision of the operands from the precision of the operations performed on them. Otherwise the proposals depend on the availability of clean, decimal, floating-point arithmetic, along with the ability to specify precisely what to do in case an exception arises.

Most of the paper has been devoted to examples which illustrate the use of these facilities. It was intended that these examples demonstrate the ease with which a very wide variety of numerical processes can be implemented with the aid of the proposed facilities. The user has a wide choice as to what he can implement, perhaps a choice that is sufficiently wide for any practical purpose.

Experience with preprocessors, the design of a complete language, and the design and construction of a special hardware unit, have provided support for the claim that the proposed language facilities can be provided both economically and efficiently.

Bibliography

[1] M. Cohen, V.C. Hamacher and T.E. Hull. CADAC: An Arithmetic Unit for Clean Decimal Arithmetic and Controlled Precision, Proceedings 5th Symposium on Computer Arithmetic, IEEE Computer Society, Ann Arbor, Michigan, 1981, pp.106-112.

[2] Austin Curley. PNCL: A Prototype Numerical Computation Language, M.Sc. thesis, Department of Computer Science, University of Toronto, in preparation.

[3] J.J. Dongarra, J.R. Bunch, C.B. Moler and G.W. Stewart. LINPACK Users' Guide, SIAM, Philadelphia, 1979.

[4] T.E. Hull. Semantics of Floating Point Arithmetic and Elementary Functions, Portability of Numerical Software, edited by Wayne Cowell, Springer-Verlag, 1977, pp.37-48.

[5] T.E. Hull. Desirable Floating-Point and Elementary Functions for Numerical Computation, Proceedings Conference on the Programming Environment for Development of Numerical Software, SIGNUM Newsletter 14, 96-99, 1979, and a similar paper in Proceedings 4th Symposium on Computer Arithmetic, IEEE Computer Society, Santa Monica, California, 1978, pp.63-69.

[6] T.E. Hull and J.J. Hofbauer. Language Facilities for Multiple Precision Floating Point Computation, with Examples, and the Description of a Preprocessor, Technical Report 63, Department of Computer Science, University of Toronto, 1974.

[7] T.E. Hull and J.J. Hofbauer. Language Facilities for Numerical Computation, Proceedings ACM-SIAM Conference on Mathematical Software II, Purdue University, 1974, pp.1-18.

GENERALIZED HERMITIAN MATRICES:

A NEW FRONTIER FOR NUMERICAL ANALYSIS?

Peter Lancaster

1. Introduction and prologue

In this expository review some ideas and results of linear algebra, both old and new, are presented. The rationale for doing so in the context of this conference is the conviction that understanding of the linear algebra is a prerequisite to good numerical linear algebra. A number of recent theoretical results will be included, but always without proof. It is our objective to give only the flavour of the theory and results.

Applications to two specific problem areas will be described, both of which are of current interest in engineering. Algorithms and computation are being reported in the engineering literature on these problems, but the involvement of professional numerical analysts is still quite small: hence the phrase "new frontier" in the title. A small sample of some papers on computation in these two areas is contained in references [16] to [20]. We remark that there are also several other areas of application to which the theory described herein is relevant.

It will be seen that invariant subspaces of matrices play an important part in our discussions, and the stability of solutions to the two problems on which we focus will be characterized by a notion of stability for invariant subspaces. Since numerical solution of these problems may be accomplished by the numerical determination of invariant subspaces, and since these concepts are also of wider interest, we introduce them briefly in this prologue.

If A is an $n\times n$ complex matrix (also seen as a transformation on the space $\mathbb{C}^n$ of column n-vectors) a subspace S of $\mathbb{C}^n$ is said to be *invariant* under A, or an A-invariant subspace if for every $x \in S$, $Ax \in S$; we also write $AS \subset S$. It is well-known that bases for an A-invariant subspace can be constructed given complete information on the eigenvectors and generalized eigenvectors of A. (For example, eigenvectors generate invariant subspaces.) An intimately connected notion is that of the Jordan normal form for A which displays not only the eigenvalues of A, but also the lengths of associated chains of generalized eigenvectors. Now Golub and Wilkinson [8] do not advise the computation of complete Jordan structure for a matrix but, nevertheless, computation of a single invariant subspace may be feasible and even efficient, if one computes an orthonormal basis for the subspace rather than a basis of eigenvectors and generalized eigenvectors.

Next we note that, in common with the Jordan normal form, an invariant subspace may be unstable under perturbations of the matrix; i.e. small perturbations in the matrix coefficients may lead to wild variations in some invariant subspaces. For this reason, it seems to be important to recognize those invariant subspaces which will be stable under perturbations and, furthermore, these are likely to be the invariant subspaces of physical significance in applications. For a formal definition we have: an A-invariant subspace S is *stable* if, given $\varepsilon > 0$, there is a δ such that $\|B-A\| < \delta$ implies that matrix B has an invariant subspace T such that $\rho(S,T) < \varepsilon$. (Here, $\|\cdot\|$ denotes a matrix-norm and ρ is a metric on the set of subspaces of $\mathbb{C}^n$, respectively. Here and elsewhere we will avoid technicalities.)

We complete this prologue by noting that all such stable invariant subspaces have been described by Bart, Gohberg and Kaashoek in chapter VIII of [2].

THEOREM 1. *Let matrix A have distinct eigenvalues $\{\lambda_1,\ldots,\lambda_r\}$. An A-invariant subspace S is stable if and only if S is a direct sum of invariant subspaces $S_1,\ldots,S_r$ where each S_k is one of:*

(a) $\{0\}$; *a trivial invariant subspace;*

(b) *a complete generalized eigenspace of λ_k, say G_k;*

(c) *any A-invariant subspace of G_k, provided the eigenspace of λ_k has dimension one.*

Note the "eigenspace of λ_k" is spanned by the eigenvectors of λ_k and the "complete generalized eigenspace of λ_k" is spanned by all eigenvectors *and* generalized eigenvectors of λ_k.

To illustrate; the 2×2 zero matrix has no one-dimensional stable invariant subspace. In contrast all invariant subspaces of $\begin{bmatrix} 0 & 1 \\ 0 & 0 \end{bmatrix}$ are stable. More generally, *all* invariant subspaces of a companion matrix (or a non-derogatory matrix) are stable. This is because, as is well-known, every distinct eigenvalue of such a matrix has an eigenspace of dimension one so that clause (c) of the theorem will always apply. (A stability result in which companion matrices arise in a similar way appears in the work of Arnold [1].)

We proceed now to the main topic of this paper.

2. Generalized hermitian matrices

If x^* denotes the conjugate transpose of a vector x in $\mathbb{C}^n$, the usual scalar product (,) is defined by

$$(x,y) = x^*y, \qquad \text{for all } x,y \in \mathbb{C}^n .$$

Then $(x,x)^{\frac{1}{2}}$ is the euclidean norm of x and x,y are orthogonal, in the euclidean sense, if $(x,y) = 0$. We first want a more general notion of orthogonality. If $H \in \mathbb{C}^{n\times n}$, $H^* = H$ and $\det H \neq 0$ we define an *indefinite* scalar product $[.,.]$ on $\mathbb{C}^n$ by

$$[x,y] = (x,Hy) = x^*Hy, \qquad \text{for all } x,y \in \mathbb{C}^n .$$

Since $H^* = H$, $[x,x]$ must be real for any x but, in general, it may be positive, zero, or negative and so $[\ ,\ ]$ does not define a norm on $\mathbb{C}^n$. However, it is useful to retain some language of geometry and say that x,y are *orthogonal* in the indefinite scalar product if $[x,y] = x^*Hy = 0$; or one may describe x and y as "H-orthogonal".

More generally, if S is a subspace of $\mathbb{C}^n$ we use $S^\perp$ for the set of all vectors in $\mathbb{C}^n$ orthogonal to S *in this new sense*. Thus,

$$S^\perp = \{x \in \mathbb{C}^n \colon [x,y] = 0 \quad \text{for all } y \in S\} .$$

In general, $S^\perp$ is not *complementary* to S, for there may be a non-zero $x \in S$ with $[x,x] = 0$ in which case we may have $x \in S^\perp$ also. Subspaces for which $S \cap S^\perp = \{0\}$ are called non-degenerate. Then a simple lemma is useful:

LEMMA. (a) *S non-degenerate implies $\mathbb{C}^n = S \oplus S^\perp$*
(b) *$\mathbb{C}^n = S \oplus S^\perp$ implies that S is non-degenerate.*

In contrast, a subspace S is said to be *neutral* (or isotropic) if $[x,x] = 0$ for all $x \in S$.

As a small illustration take $H = \begin{bmatrix} 0 & 1 \\ 1 & 0 \end{bmatrix}$ and S the span of the first unit vector. It is easily seen that S is neutral in the indefinite scalar product defined by H. Furthermore, $S^\perp = S$.

So much for geometry; now we say that, with H and $[\ ,\ ]$ as above, a matrix A is *hermitian with respect to* $[\ ,\ ]$, *or to* H, if

$$[x,Ay] = [Ax,y] \qquad \text{for all } \ x,y \in \mathbb{C}^n.$$

This definition is clearly equivalent to

$$(x,HAy) = (Ax,Hy), \qquad \text{for all } \ x,y \in \mathbb{C}^n,$$

and to

$$HA = A^*H, \quad \text{or} \quad A^* = HAH^{-1} . \tag{1}$$

Thus, the last statement says A is hermitian with respect to H if and only if A^* is similar to A and H defines the similarity. Note that, if $H = I$, then obviously $A^* = A$, and $[\ ,\ ] = (\ ,\ )$.

It should be emphasized that this generalization includes matrices arising in many important applications. They include Hamiltonian and symplectic matrices and arise in several branches of engineering and theoretical physics.

Our next idea is to see to what extent the nice spectral properties of hermitian matrices carry over to this weaker form of symmetry. Some answers are summarized in the next lemma.

Recall that here, as in the prologue, G_k denotes a complete generalized eigenspace associated with one of the distinct eigenvalues λ_k of A.

LEMMA. *If A is hermitian with respect to H, then*

(i) $\lambda_j \neq \overline{\lambda_k}$ *(λ_j, λ_k not a conjugate pair) implies* $G_j \subset G_k^{\perp}$.

(ii) $\lambda_j \neq \overline{\lambda_j}$ *(λ_j non-real) implies* G_j *is neutral.*

(iii) $AS \subset S$ *implies* $AS^{\perp} \subset S^{\perp}$.

Item (ii) here is, in fact, a phenomenon not to be anticipated from the theory of hermitian matrices. Items (i) and (iii), on the other hand, seem quite familiar and natural. Thus, we have a weak form of orthogonality for the eigenspaces which reduces to euclidean orthogonality when $H = I$. In the latter case, this property makes computation with hermitian matrices a relatively simple and well-understood task. The question mark in the title of this paper relates more specifically to the question of whether, in computation with our more general hermitian matrices, some advantage can be taken of the weaker form of orthogonality?

The discussion in this section is, of course, quite widely known and understood. A comprehensive and more general treatment can be found in the work of Bognar [3].

Before discussing two interesting applications it is necessary to present a less familiar theorem. Indications of this result can be discovered in the work of Malcev [13], Thompson [15] and others, but first appeared in the present form in [6]. It concerns the simultaneous reduction of A by similarity and H by congruence. To motivate the result, suppose A is hermitian with respect to H, then we have seen in (1) that $A^* = HAH^{-1}$, so A and A^* have the same eigenvalues, and this implies that the non-real eigenvalues of A must appear in conjugate pairs. It follows that, if J is a Jordan normal form for A, it can be partitioned into block-diagonal form: $J = \mathrm{diag}\{J_c, J_r, \overline{J_c}\}$ where J_r is real and J_c has no real eigenvalues and no conjugate complex pairs of eigenvalues. Also, there is a T for which $A = TJT^{-1}$.

Now if $H = I$ we would have $A^* = A$ and, *in addition*, we could choose $T^*T = I$. More generally (when $H \neq I$), what can be said about the transforming matrix T? Loosely, there is a canonical matrix $P_{\varepsilon,J}$, defined by A and H (whose elements are all 0, +1 or -1) so that we have:

THEOREM 2. *Matrix A is hermitian with respect to H if and only if there is an invertible T such that*

$$T^{-1}AT = J \quad \text{and} \quad T^*HT = P_{\varepsilon,J}.$$

We will not even define $P_{\varepsilon,J}$ exactly, but refer the reader to [6] or [7] for details. However, we indicate the following nice properties. When partitioned in conformity with $\mathrm{diag}\{J_c, J_r, \overline{J_c}\}$,

$$P_{\varepsilon,J} = \begin{bmatrix} 0 & 0 & P_c \\ 0 & P_r & 0 \\ P_c & 0 & 0 \end{bmatrix},$$

and also (writing P for brevity), $P^* = P$, $PJ = J^*P$, $P^2 = I$. The subscript ε denotes a set of signs appearing in P_r (the A-sign characteristic of H) and the subscript J emphasizes the very strong dependence of P on J. Also, when $H = I$, P reduces to I and then T is unitary.

The proof of the theorem relies heavily on the preceding lemmas to cut down the action of A to successively smaller A-invariant subspaces.

An important corollary (the result obtained in [15]) concerns the simultaneous reduction of two quadratic forms, neither of which is positive definite.

COROLLARY. *If* $B = B^*$, $H = H^*$ *and* $\det H \neq 0$, *then there is a nonsingular* T *such that*

$$T^*BT = P_{\varepsilon,J}J \qquad \textit{and} \qquad T^*HT = P_{\varepsilon,J}.$$

The proof follows immediately from the theorem if we define $A = H^{-1}B$ and observe that

$$HA = B = B^* = A^*H$$

so that A is hermitian with respect to H. Applying the theorem we immediately get the second relation of the corollary and for the first:

$$T^*BT = (T^*HT)(T^{-1}H^{-1}BT) = (T^*HT)(T^{-1}AT) = P_{\varepsilon,J}J.$$

In the two following applications we make use of the theorem in the construction of invariant subspaces of A having specified sign (or neutrality) with respect to H. To illustrate how this can be done, let t_j be a column of T in the theorem which is also an eigenvector of A. Thus $AT = TJ$ will yield $At_j = t_j\lambda_j$ for some eigenvalue λ_j and the span of t_j is an A-invariant subspace, call it S. But the second result tells us that

$$[t_j, t_j] = t_j^*Ht_j = 0, \text{ or } 1, \text{ or } -1.$$

Thus, depending on the jth diagonal element of $P_{\varepsilon,J}$ we can assert that S is H-neutral, or H-positive, or H-negative.

3. The solutions of an algebraic Riccati equation

As our first application consider the classical optimal control problem for the system

$$\frac{dx}{dt} = -Ax + Bu, \qquad y = Hx$$

(A, B, H are matrices with A of size $n\times n$). A control vector u is to be found which minimizes the functional

$$\frac{1}{2}\int_0^\infty (y^*Qy + u^*Ru)dt,$$

where R is positive definite and $Q^* = Q$. The solution is due to Kalman and asserts that, given a controllability condition, the optimal control is $u = -R^{-1}B^*Xx$ where $X = X^*$ is a hermitian solution of

$$XDX + XA + A^*X - C = 0 \tag{2}$$

where $D = BR^{-1}B^*$ (and is clearly positive semidefinite) and $C = H^*QH$. The controllability condition is that

$$\mathrm{rank}[D\ AD\ \dots\ A^{n-1}D] = n\ .$$

Equation (2) is described as a symmetric algebraic Riccati equation, and the existence and uniqueness of hermitian solutions X has been investigated by Lancaster and Rodman in [10]. In those investigations considerable use was made of the results described in Section 2. To see how this might come about we define

$$M = \begin{bmatrix} A & D \\ C & -A^* \end{bmatrix}, \qquad H = \begin{bmatrix} -C & A^* \\ A & D \end{bmatrix}$$

and observe that $H^* = H$. Furthermore, it can be assumed without loss of generality that H is invertible. Then it is easily verified that $H(iM) = (iM)^*H$, so that iM is hermitian with respect to H.

Now we make the following simple observations which immediately suggest that Theorem 2 may have some usefulness here.

LEMMA. *If there is a solution $X = X^*$ of equation* (2) *then the range of $\begin{bmatrix} I \\ X \end{bmatrix}$ is M-invariant and H-neutral.*

<u>Proof</u>. Observe that

$$M\begin{bmatrix} I \\ X \end{bmatrix} = \begin{bmatrix} A & D \\ C & -A^* \end{bmatrix}\begin{bmatrix} I \\ X \end{bmatrix} = \begin{bmatrix} A+DX \\ C-A^*X \end{bmatrix} = \begin{bmatrix} A+DX \\ XDX+XA \end{bmatrix} = \begin{bmatrix} I \\ X \end{bmatrix}(A+DX),$$

and the first assertion follows immediately.

Then multiply the matrix product out to see that

$$[I\ X^*]\ H\begin{bmatrix} I \\ X \end{bmatrix} = [I\ X^*]\begin{bmatrix} -C & A^* \\ A & D \end{bmatrix}\begin{bmatrix} I \\ X \end{bmatrix} = 0\ ,$$

from which the lemma follows immediately.

Using the canonical forms of Theorem 2 a converse statement can also be obtained to arrive at:

THEOREM 3. *If D is semidefinite and (A,D) controllable then the following are equivalent*:

(i) *there exists a solution* $X = X^*$ *of* (2).

(ii) *there exists an* M*-invariant subspace* S *of dimension* n *which is* H*-neutral.*

(iii) *the partial multiplicities of all real eigenvalues (if any) of* iM *are even.*

A very simple example is given by

$$D = \begin{bmatrix} 1 & 0 \\ 0 & 0 \end{bmatrix}, \qquad A = \begin{bmatrix} 0 & 0 \\ 1 & 0 \end{bmatrix}, \qquad C = \begin{bmatrix} -2 & 0 \\ 0 & 1 \end{bmatrix}.$$

It is found that iM has eigenvalues +1 and -1 and the elementary divisors of $I\lambda - iM$ are $(\lambda-1)^2$ and $(\lambda+1)^2$. So there is a hermitian solution. In fact, $X = \begin{bmatrix} 0 & -1 \\ -1 & 0 \end{bmatrix}$.

If C is replaced by $-C$ it is found that M has two pure imaginary eigenvalues and two real eigenvalues each of multiplicity one; so the Riccati equation has no hermitian solution.

In general, there are several hermitian solutions to (2) and, as is well-known in the field, the choice of solution influences the location of the eigenvalues of $A + DX$ in the complex plane. The theory described here admits the proof of:

COROLLARY. *There is a solution* $X = X^*$ *and* $A + DX$ *is strictly stable, i.e.*

$$\sup_{t\geq 0} \|e^{(A+DX)t}\| < \infty ,$$

if and only if the partial multiplicities of all real eigenvalues of iM *(if any) are exactly* 2.

When the corollary applies the solution X described is unique and is called the *maximal* solution.

4. The factorization of matrix polynomials

In this section we indicate how the results of section 2 apply to the factorization of matrix polynomials. These problems are of considerable interest in system theory and filtering of multi-channel time series, as well as other areas of application. The presentation is based primarily on [6] and [7] and generalizations to rational matrix functions can be found in [2], (see also [20]).

We shall consider only monic matrix polynomials (m.m.p.), i.e. matrix valued functions of the form

$$L(\lambda) = \sum_{j=0}^{\ell} A_j \lambda^j, \qquad A_\ell = I$$

where $A_0, A_1, \ldots, A_\ell$ are complex $n \times n$ matrices. The problem is to find m.m.p. L_1 and L_2 for which $L = L_2 L_1$.

We first note such polynomials have coefficients in a ring and not a field so the fundamental theorem of algebra does not apply. There may be *no* such factorization possible. For example, there are no 2×2 matrices A and B for which

$$L(\lambda) = \begin{bmatrix} 1 & 0 \\ 0 & 1 \end{bmatrix}\lambda^2 + \begin{bmatrix} 0 & 1 \\ 0 & 0 \end{bmatrix} = (I\lambda+A)(I\lambda+B) \ . \tag{3}$$

The first important result to be described characterizes, in a geometrical way, those situations where a factorization exists. Since the proof, and even a full statement of the theorem, involve technicalities our treatment will be superficial. Full details can be found in [4] and [7].

"THEOREM" 4. *A non-trivial factorization* $L = L_2L_1$ *(where* L, L_1, L_2 *are m.m.p.) is characterized by the existence of a certain invariant subspace of the* $\ell n\times\ell n$ *companion matrix*

$$C_L = \begin{bmatrix} 0 & I & 0 & \cdots & 0 \\ 0 & 0 & I & & 0 \\ \vdots & & & \ddots & \vdots \\ 0 & 0 & 0 & \cdots & I \\ -A_0 & -A_1 & -A_2 & \cdots & -A_{\ell-1} \end{bmatrix} .$$

An invariant subspace of C_L with the special properties required is called a *supporting subspace* for the associated right divisor, L_1. The 2×2 example above gives a negative illustration. The 4×4 companion matrix has *no* 2-dimensional invariant subspace with the special property required of a supporting subspace.

The theory of section 2 becomes relevant in the case of a *selfadjoint* m.m.p. $L(\lambda) = \sum_{j=0}^{\ell} A_j\lambda^j$, i.e. one for which $A_j^* = A_j$, $j = 0,1,\ldots,\ell$, $A_\ell = I$. In this case, we write $L^* = L$. First observe that if B is the matrix with block Hankel form:

$$B = \begin{bmatrix} A_{\ell-1} & A_{\ell-2} & \cdots & A_1 & I \\ A_{\ell-2} & & & I & 0 \\ \vdots & & \iddots & & \\ A_1 & I & & & \\ I & 0 & & & 0 \end{bmatrix}$$

then $B^* = B$ and B is nonsingular. Furthermore, a little calculation shows that $BC_L = C_L^*B$. So C_L *is hermitian with respect to* B and we are in business!

It is now possible to make heavy use of theorem 2 and construct C_L-invariant subspaces which are nonnegative with respect to B. Such subspaces of the largest possible dimension turn out to have also the special property required to be supporting, and so, by theorem 4, define a right divisor L_1. The degree of L_1 is determined by the dimension of the supporting subspace. In this way, a proof of the following existence theorem (due to Langer [12] but see also [6]) is obtained.

THEOREM 5. *Let L be a selfadjoint m.m.p. of degree ℓ.*

(a) *If $\ell = 2k$ then there is a m.m.p. L_1 of degree k for which $L = L_2L_1$ and the supporting subspace S for L_1 is C_L-invariant and maximal B-nonnegative.*

(b) *If $\ell = 2k+1$ the conclusions of* (a) *also apply for a right divisor L_1 of degree k.*

Note that, in contrast to the negative example above, a *selfadjoint* quadratic m.m.p.

$$L(\lambda) = I\lambda^2 + A_1\lambda + A_0, \quad A_1^* = A_1, \quad A_0^* = A_0$$

can always be factored in the form $L(\lambda) = (I\lambda+A)(I\lambda+B)$. The reader may have surmised, correctly, that this elegant result is remarkably difficult to prove. It originates with Langer [11] (in greater generality); see also Krein and Langer [9].

5. Stability

We conclude this paper with some remarks on the question of stability and this will admit a return to the notion introduced in the prologue. We begin with the topic of factorization of m.m.p., and ask when a factorization is stable under perturbation of the coefficients of the parent m.m.p..

When $n > 1$ (L is $n\times n$) new phenomena arise. For example, if

$$L(\lambda,\mu) = I\lambda^2 + \begin{bmatrix} 0 & \mu \\ 0 & 0 \end{bmatrix}, \tag{4}$$

then, as in the case of (3) there is *no* factorization of the form $(I\lambda+A(\mu))(I\lambda+B(\mu))$ as long as $\mu \neq 0$. However there is an obvious factorization when $\mu = 0$. We would say that $L(\lambda,\mu)$ has an *isolated* divisor at $\mu = 0$. Likewise, if

$$L(\lambda,\mu) = I\lambda^2 + \begin{bmatrix} 0 & \mu \\ 1 & 0 \end{bmatrix}$$

then $L(\lambda,\mu)$ *has* a divisor at all points of a deleted neighbourhood of $\mu = 0$ but not at $\mu = 0$ itself. Such questions of parameter dependence are considered in [5].

Here, we will consider a more general form of perturbation. First introduce a norm on the m.m.p.: $\|L\| = \sum_{j=0}^{\ell} \|A_j\|$. Then we say that a factorization $L = L_2L_1$ in m.m.p. is stable if, given $\varepsilon > 0$ there is a δ such that, for any m.m.p. with $\|L-\hat{L}\| < \delta$, there is a right divisor $\hat{L}_1$ of $\hat{L}$ for which $\|L_1-\hat{L}_1\| < \varepsilon$. Thus, the factorization of (4) at $\mu = 0$ is not stable.

Now by theorem 4 we can equate the existence of divisors with the existence of a supporting subspace and then, referring to theorem 1, enquire about the relationship between stability of the divisor and the supporting subspace. It turns out that they are *equivalent*. Using this equivalence Bart, Gohberg and Kaashoek (chapter VIII of [2]) were able to characterize stable factorizations as follows:

THEOREM 6. *The factorization* $L = L_2L_1$ *(in m.m.p.) is stable if and only if, for each eigenvalue* λ_0 *common to* L_2 *and* L_1, *the dimension of the eigenspace of* $L(\lambda_0)$ *is one.*

Note that eigenvalues of L are the zeros of det $L(\lambda)$ and the eigenspace of $L(\lambda_0)$ is the nullspace, or kernel, of $L(\lambda_0)$. In particular, if there is a factorization $L = L_2L_1$ *and* L_2 and L_1 have no common eigenvalues then the theorem implies the stability of the factorization. In this case, the factorization is said to be *spectral* and the stability is, for other reasons, not surprising.

Note also that when $n = 1$, L is just a scalar complex polynomial and the theorem implies that all factorizations are stable, as is otherwise obvious.

Finally, we comment on the stability of solutions of the Riccati equation (2). First a solution X of (2) is said to be stable if, given $\varepsilon > 0$, there is a δ such that

$$\max(\|\hat{A}-A\|, \|\hat{D}-D\|, \|\hat{C}-C\|) < \delta$$

implies that $X\hat{D}X + X\hat{A} + \hat{A}^*X - \hat{C} = 0$ has a solution $\hat{X}$ such that $\|\hat{X}-X\| < \varepsilon$.

Using theorem 3 now, stability of a solution can be related to the stability of an invariant subspace of M and theorem 1, once again, gives a "handle" on this problem. We remark only that the parameter dependence of the maximal solution of (2) has been investigated by Rodman [14], and that investigations of the stability of solutions of the more general non-symmetric Riccati equation have been undertaken in [2].

References

[1] V.I. Arnold *On matrices depending on parameters* Russian Math. Surveys (translation) 26 (1971), 29-43.

[2] H. Bart, I. Gohberg and M.A. Kaashoek *Minimal Factorization of Matrix and Operator Functions* Birkhäuser Verlag, Basel, Boston, Stuttgart, 1979.

[3] J. Bognar *Indefinite Inner Product Spaces* Springer-Verlag, New York, Heidelberg, Berlin, 1974.

[4] I. Gohberg, P. Lancaster and L. Rodman *Spectral analysis of matrix polynomials - I. Canonical forms and divisors* Lin. Alg. & Appl. 20 (1978), 1-44.

[5] I. Gohberg, P. Lancaster and L. Rodman *Perturbation theory for divisors of operator polynomials* SIAM J. Math. Anal. 10 (1979), 1161-1183.

[6] I. Gohberg, P. Lancaster and L. Rodman *Spectral analysis of selfadjoint matrix polynomials* Annals of Math. 112 (1980), 33-71.

[7] I. Gohberg, P. Lancaster and L. Rodman *Matrix Polynomials* Academic Press (to appear).

[8] G. Golub and J.H. Wilkinson *Ill-conditioned eigensystems and computation of the Jordan canonical form* SIAM Rev. 18 (1976), 578-619.

[9] M.G. Krein and H. Langer *On some mathematical principles in the linear theory of damped oscillations of continua* Int. Eqs. and Op. Theory, 1 (1978), 364-399 (Translation).

[10] P. Lancaster and L. Rodman *Existence and uniqueness theorems for the algebraic Riccati equation* Int. J. Control, 32 (1980), 285-309.

[11] H. Langer *Spektaltheorie linearer Operatoren in J-Räumen und einige Anwendungen auf die Schar* $L(\lambda) = I\lambda^2 + B\lambda + C$ Habilitationschrift Tech. Univ. Dresden, 1964.

[12] H. Langer *Factorization of operator pencils* Acta. Sci. Math. (Szeged) 38 (1976), 83-96.

[13] A.I. Mal'cev *Foundations of Linear Algebra* W.H. Freeman, San Francisco and London, 1963.

[14] L. Rodman *On extremal solutions of the algebraic Riccati equation* Lectures in Appl. Math. 18 (1980), 311-327.

[15] R.C. Thompson *The characteristic polynomial of a principal subpencil of a Hermitian matrix pencil* Lin. Alg. & Appl. 14 (1976), 135-177.

References on Numerical Techniques

[16] D.L. Kleinman *On an iterative technique for Riccati equation computation* IEEE Trans. Autom. Control, AC-13 (1968), 114-115.

[17] A.J. Laub *A Schur method for solving algebraic Riccati equations* Proc. 1978 IEEE Trans. Autom. Control AC-24 (Dec. 1979), 913-921.

[18] C. Paige and C. Van Loan *Orthogonal reductions of the algebraic Riccati equation* Cornell Univ., Computer Sc. Tech. Rep. TR 79-377, 1979.

[19] T. Pappas, A.J. Laub and N.R. Sandell, Jr. *On the numerical solution of the discrete-time algebraic Riccati equation* IEEE Trans. Autom. Control, AC-25 (1980), 631-641.

[20] P.M. Van Dooren and P. DeWilde *Minimal cascade factorization of real and complex rational transfer matrices* IEEE Trans. Circuits and Systems, CAS-28 (1981), 390-400.

Some Applications of Geometry In Numerical Analysis

Robin J. Y. McLeod

Abstract. Surface interpolation is used to produce basis functions for two-dimensional curved finite elements. The connection between rational surfaces and isoparametric methods is discussed with the occurrence of the Steiner surface being highlighted. A corollary to Max Noether's intersection theorem is used to produce high order stable bases. Parametric cubic curves are discussed from a geometrical viewpoint and this viewpoint is used to develop transfinite blending functions for a wide variety of shapes.

Introduction.

There are many more areas of application of geometry in numerical analysis than are discussed in this paper. The choice made reflects both limitations of space and the authors own research interests. It is hoped that these discussions will illustrate that in at least the areas mentioned a geometric viewpoint has not only given a new and powerful insight into a well known problem but also produced many new results. So many of the finite element interpolation problems are seen to be special cases of more general geometric results. Potential problems in parametric surface design or approximation are highlighted from the study of even a simple rational surface and important properties of all types of parametric cubic curve approximation are transparent when one studies the rational cubic curve. The underlying theme is that of interpolation, a very old and very pervasive branch of numerical analysis.

1. A Finite Element Basis Function As A Surface

Consider the problem of defining a basis function for a two dimensional finite element. We will take x and y as the usual cartesian coordinates in the plane. Let the basis function be labelled z. Hence we seek some $z(x,y)$ which satisfies certain properties. As a simple first example consider the situation of figure 1a where, without loss of generality, we have assumed that a linear transformation has been done so that the vertices of the

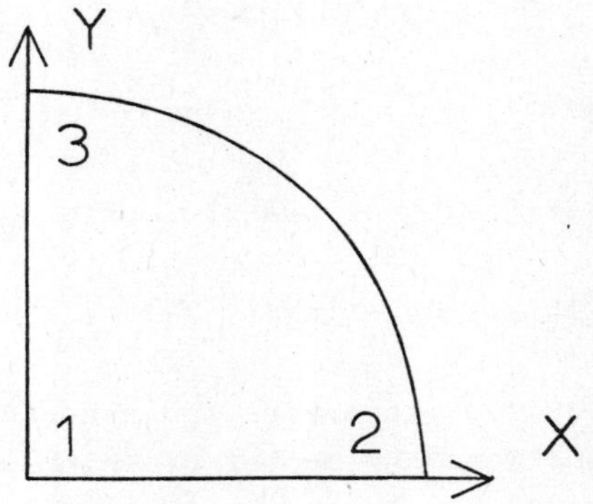

Figure 1a.

three-sided element are $(0,0)$, $(1,0)$ and $(0,1)$ numbered 1, 2 and 3 respectively. Sides 12 and 13 are straight and side 23 is curved with equation assumed known and given by

$$f(x,y) = 0 \quad . \tag{1}$$

A simple and typical set of conditions on the basis function corresponding to node 1 are the following:

$$z(x,0) = 1 - x \;, \qquad z(0,y) = 1 - y \tag{2}$$

$$z = 0 \;\; \text{when} \;\; f(x,y) = 0 \quad .$$

This can be viewed as a surface interpolation problem where we seek a surface given, say, by $F(x,y,z) = 0$ which passes through the three space curves given by

$$z + x - 1 = 0 \;, \quad y = 0$$

$$z + y - 1 = 0 \;, \quad x = 0 \tag{3}$$

and

$$z = 0 \;, \quad f(x,y) = 0 \quad .$$

Now it is easy to construct equations for a wide variety of surfaces satisfying conditions 3. The last of conditions 3 is satisfied by any surface of the form

$$F(x,y,z) \equiv zg(x,y,z) + f(x,y) = 0 \tag{4}$$

where $g(x,y,z)$ is any well defined function. As a simple example assume that $f(x,y) = 0$ is a conic given by

$$f(x,y) \equiv ax^2 + bxy + cy^2 - (1+a)x - (1+c)y + 1 = 0 \tag{5}$$

Then the one parameter family of quadric surfaces given by

$$\alpha z^2 + [\alpha(x+y-1) + (ax+cy-1)]z + f(x,y) = 0 \tag{6}$$

satisfies all the required properties and hence defines a single infinity of basis functions any member of which would be suitable as a member of a first order basis for the curved element of figure 1a [10]. For $\alpha = 1$ the surface is a cone and for $\alpha > 1$ the part of the surface above the element is below the cone, figures

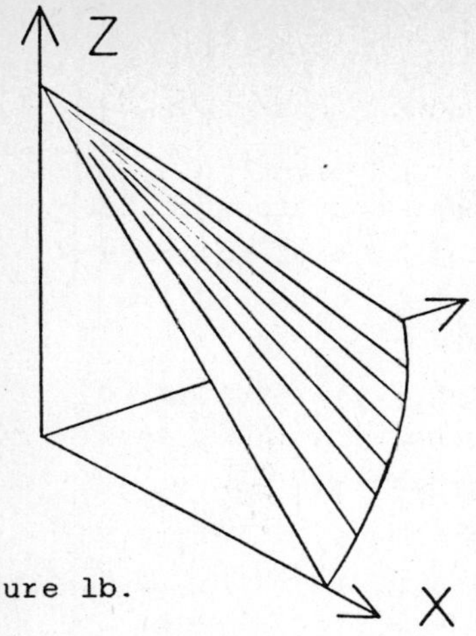

Figure 1b.

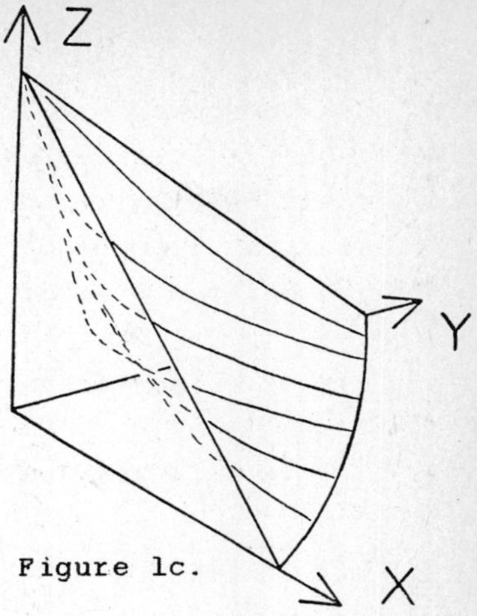

Figure 1c.

1b and 1c. For $\alpha = 0$ equation 6 gives a rational solution for z

$$z = \frac{f(x,y)}{1-ax-cy} \quad . \tag{7}$$

The concept of producing rational bases has been greatly extended beyond this simple example and we will return to this later [16].

Let us consider another example where now we would like our basis function to have quadratic variation along the adjacent element sides. The equivalent surface interpolation problem is now to find a surface passing through the space curves given by

$$\begin{aligned} z &= (1-x)(1-2x) \quad , \quad y = 0 \\ z &= (1-y)(1-2y) \quad , \quad x = 0 \\ z &= 0 \quad , \quad f(x,y) = 0 \end{aligned} \tag{8}$$

Consider $F(x,y,z)$ to be a polynomial of degree four. The satisfaction of the third of conditions 8 is achieved by choosing the terms not involving z to be of the form $q_2(x,y)\, f(x,y)$ where $q_2(x,y)$ is a polynomial of degree 2. We are assuming as before that $f(x,y) = 0$ is a conic though this restriction could easily be relaxed.

Hence the surface $F(x,y,z) = 0$ of the form

$$F(x,y,z) \equiv zq_3(x,y,z) + q_2(x,y)f(x,y) = 0 \tag{9}$$

where $q_3(x,y,z)$ is a polynomial of degree 3, satisfies the third of conditions 8. There are 25 free parameters in equation 9. Choose the coefficient of z^4 to be zero and likewise the coefficients of z^3x and z^3y. The satisfaction of the remainder of conditions 8 gives an additional 13 linear restrictions on the coefficients in equation 9. There still remains a 9 parameter family of quartic surfaces all satisfying the interpolatory conditions and hence all providing basis functions with suitable properties on the element boundary. A special case of equation 9 satisfying condition 8 is given by

$$\alpha^2\beta^2z^2 + z[(3\alpha\beta+4\beta-4\alpha)(\alpha^2y+\beta^2x) + 3\alpha^2\beta^2(x+y) - 2(\alpha^2+\beta^2)$$

$$+ \alpha\beta(4-2\alpha\beta+3\alpha-3\beta) - 2(\alpha^2+\beta^2)(\alpha y+\beta x)^2]$$

$$+ [\beta x+\alpha y)^2 + (-\alpha\beta-\beta+\alpha)(x+y-1) - (\alpha^2y+\beta^2x)][4(\alpha y+\beta x)^2$$

$$+ (\alpha\beta+2\beta-2\alpha)(1-2x-2y) - 2(\alpha^2y+\beta^2x)] = 0 \quad . \tag{10}$$

This particular quartic surface is rational having parametization given by

$$x = p(1=\alpha q) \qquad y = q(1+\beta p) \qquad z = (1-p-q)(1-2p-2q) \tag{11}$$

This parameterization is usually obtained directly from the quadratic isoparametric transformation using the usual quadratic polynomials on a triangle to define the transformation

$$x = \sum_{i=1}^{6} x_iW_i(p,q) \qquad y = \sum_{i=1}^{6} y_iW_i(p,q) \tag{12}$$

where the polynomials $W_i(p,q)$ $i = 1, 2, \ldots, 6$ are given by

$$\begin{aligned}
W_1(p,q) &= (1-p-q)(1-2p-2q)\\
W_2(p,q) &= 4p(1-p-q)\\
W_3(p,q) &= p(2p-1)\\
W_4(p,q) &= 4pq\\
W_5(p,q) &= q(2q-1)\\
W_6(p,q) &= 4q(1-p-q)
\end{aligned} \tag{13}$$

the (x_i, y_i) $i = 1, 2, \ldots, 6$ are $(0,0)$, $(\frac{1}{2}, 0)(1,0)$, $(\frac{1}{2}, \frac{1}{2})$, $(0,1)$ and $(0, \frac{1}{2})$ respectively and $\alpha = 2(2x_4-1)$, $\beta = 2(2y_4-1)$. In this case the transformation 14 maps the unit triangle onto one

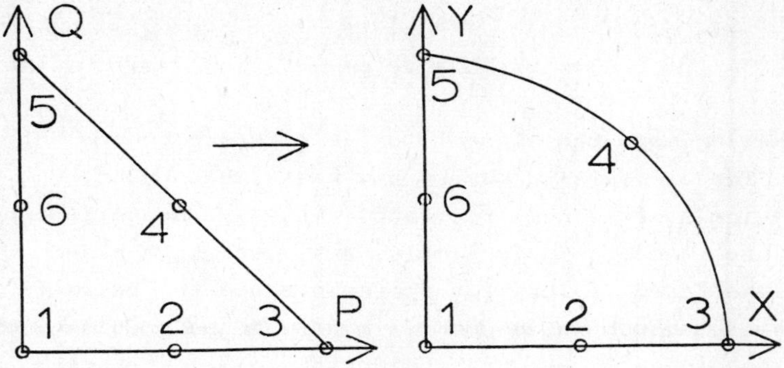

Figure 2a.   Figure 2b.

with a single curved side (figure 2). The curve is a segment of the parabola which is the image of the line $1-p-q = 0$ being given by the parameterisation

$$x = p(1+\alpha(1-p)) \qquad y = (1-p)(1+\beta p) \qquad p \in [0,1] \qquad (14)$$

The basis function associated with node 1 which we have labelled $z(x,y)$ is usually taken to be the $W_1(p,q)$ of equation 3. Thus

$$z(x,y) = W_1(p(x,y), q(x,y)) \qquad (15)$$

under the assumption that the inverse transformation of equations 12 exists. This common basis function defined with the local support of one element is indeed a part of the rational quartic given by equation 10. The key to the popularity of this basis function is the simplicity of the parameterisation rather than any particularly appealing properties of the associated quartic surface. The two however are equivalent. We continue the discussion of this surface and its parameterisation in the next section.

2. Rational Surfaces and Isoparametric Transformations

The isoparametric transformation method is defined in the following way. Let $\{\phi_i(\underline{p})\}$ $i = 1, \ldots, N$ be some basis associated with an element E, usually two or three dimensional, $\underline{p}$ being either (p,q) or (p,q,r) as the case may be. One then selects as set of coordintes $\underline{x}_i$ ie (x_i,y_i) or (x_i,y_i,z_i) i = 1, 2, ..., N and defines the transformation

$$\underline{x} = \sum_{i=1}^{N} \underline{x}_i \phi_i(\underline{p}) \qquad (16)$$

This transformation maps the element E in $\underline{p}$ space to one F, say in $\underline{x}$ space. The basis $\{\phi_i\}$ may be any basis whatsoever, polynomial, rational, trigonometric, etc. Usually the philosophy dictates that the element E is simple, say a straight sided element, or plane faced if in three dimensions and the basis is as simple as possible though we do not claim here to be precise in the use of the word "simple." Let us consider the simplest three dimensional case, that of the tetrahedron , figure 3a. The usual polynomial basis will when used in conjunction with the transformation 16 map this tetrahedron onto the element bounded by 4 curved faces, figure 3b. We consider the question of determining what these faces are. Looking at the image of a single face, that containing the 6 nodes 1,2, ..., 6 we can take the usual polynomial basis as polynomials in p,q and r. Then set 1-p-q-r = 0 to obtain the image of that plane. This is equivalent to using only the 6 basis functions given by equations 13 and mapping the element of figure 3c (which is the same as than in 2a) to a curved surface, as in figure 3d, using equation 16 viz

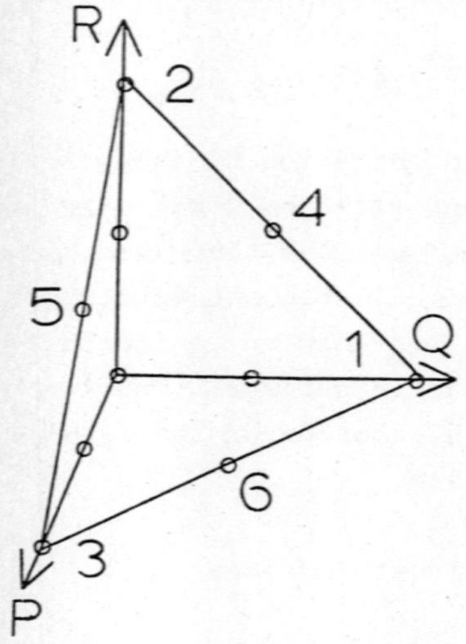

Figure 3a

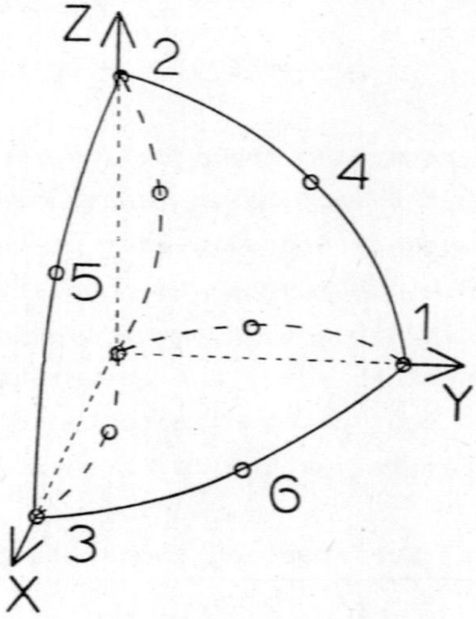

Figure 3b

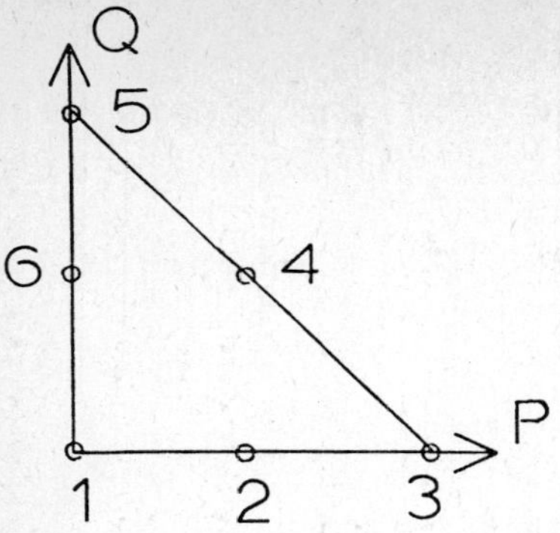

Figure 3c

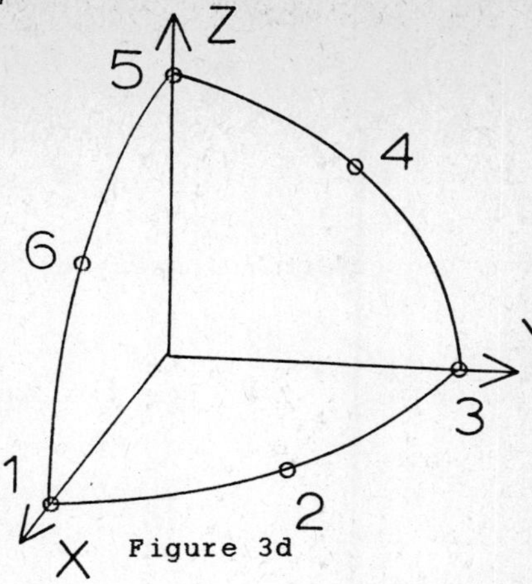

Figure 3d

$$x = \Sigma_{i=1}^{6} x_i W_i(p,q)$$

$$y = \Sigma_{i=1}^{6} y_i W_i(p,q) \qquad (17)$$

$$z = \Sigma_{i=1}^{6} z_i W_i(p,q)$$

If we choose the coordinates of nodes 1,2, ..., 6 as (1,0,0) $(x_2,y_2,0)$ (0,1,0) $(0,y_4,z_4)$, (0,0,1) and $(x_6,0,z_6)$ equations 17 give

$$x = (1-p-q)(1+2(2x_2-1)p + 2(2x_6-1)q)$$

$$y = p(1+2(2y_2-1)(1-p-q) + 2(2y_4-1)q) \qquad (18)$$

$$z = q(1+2(2z_6-1)(1-p-q) + 2(2z_4-1)q)$$

This represents another surface whose parameterisation, like that of equation 11 is given by polynomials of degree 2. These two surfaces are indeed closely related both being examples of the Steiner surface.

Making a change of variable and introducing homogeneous coordinates in the plane and in space set

$$1-p-q = \frac{u}{r}, \quad p = \frac{v}{r} \quad \text{and} \quad w = r-u-v \qquad (19)$$

$$\text{and} \quad X = \frac{x}{w}, \; Y = \frac{y}{w}, \; Z = \frac{z}{w} \qquad (20)$$

Further set

$$x_2 = y_2 = y_4 = z_4 = x_6 = z_6 = \tfrac{1}{4}(2\gamma+1) \qquad (21)$$

and equations 18 becomes

$$X = p(p+2\gamma q+2\gamma r)$$

$$Y = q(2\gamma p+q+2\gamma r) \qquad (22)$$
$$Z = r(2\gamma p+2rq+r)$$
$$W = (p+q+r)^2$$

Using homogeneous coordinates for the surface given by equation 11 gives

$$X = p(r+\alpha q)$$
$$Y = q(r+\beta p) \qquad (23)$$
$$Z = (r-p-q)(r-2p-2q)$$
$$W = r^2$$

and the similarity is stronger though perhaps not yet transparent.

Now let us consider a quartic surface having three double lines meeting in a triple point. With the origin as the triple point and the coordinate axes as the double lines. We can write this as

$$x^2y^2 + y^2z^2 + z^2x^2 - 2xyz = 0 \qquad (24)$$

Putting $x = at, \ y = bt, \ z = ct$ (25)

then $p = bc, \ q = ca, \ r = ab$ (26)

we can obtain a parametrisation of this surface in the form

$$X= 2qr, \ Y = 2pr, \ Z = 2pq, \ W = (p^2+q^2+r^2) \qquad (27)$$

Again we have a surface which has a quadratic parameterisation.

Now given a conic Ω a point O and a variable line through O the locus of points P such that OP is cut harmonically by the conic is called the polar of O with respect to the conic. O is called the pole of this line with respect to the conic. If a triangle is so placed with respect to a conic such that each of the sides is the polar of the opposite vertex with respect to the conic the triangle is said to be a self polar triangle (or self conjugate) with respect to the conic. Now if a triangle inscribed in one conic Ω' say is self polar with respect to a second conic Ω we say that Ω' is outpolar to Ω and Ω is inpolar to Ω'. It can be shown that this condition is a single linear constraint on Ω given Ω'.

Now let us return to the discussion of the surfaces given by equations 11 and 18, and that given by equation 27. All these surfaces are special cases of the one given parametrically by

$$X = C_1(p,q,r)$$

$$Y = C_2(p,q,r)$$

$$Z = C_3(p,q,r) \qquad (28)$$

$$W = C_4(p,q,r)$$

where C_1, C_2, C_3 and C_4 are homogeneous polynomials of degree two. C_1 through C_4 then represent conics in the plane and by the above argument we see that there is a single parameter family of conics inpolar to the four conics associated with the polynomials C_1, C_2, C_3 and C_4 i.e. inpolar to the system

$$\lambda C_1 + \lambda_2 C_2 + \lambda_3 C_3 + \lambda_4 C_4 = 0 \qquad (29)$$

We can think of this family as $S + \mu S'$ where S and S' are any two members of the family. We can think of this family as being the family passing through four fixed points or equivalently as being tangential to four given lines. Choosing the second of these options we further realize that the square of each of the tangent lines is now a conic outpolar to the entire system. Now if one of these repeated tangent lines does not belong to the system given by equation 29 then the system is inpolar to five linearly independent conics and hence cannot be a one parameter family. This provides a contradiction and hence the square of each of these tangents must be in the system given by equation 29. There are then four linear combinations of X, Y, Z, and W defining a new coordinate system say X_1, X_2, X_3, X_4 such that the new parameterisations are perfect squares. That is, a linear transformation in space reduces the general parameterisation given by equation 28 to a simpler one where each parameterisation is a perfect square. A further linear transformation this time in the plane and mapping p,q,r to ξ_1, ξ_2, ξ_3 say, can be chosen so as to make the four tangent lines have equations

$$\xi_1 \pm \xi_2 \pm \xi_3 = 0 \qquad (30)$$

Hence the general surface given by equation 28 is equivalent, under linear transformations, to

$$
\begin{aligned}
X_1 &= (\xi_1 - \xi_2 - \xi_3)^2 \\
X_2 &= (\xi_1 - \xi_2 + \xi_3)^2 \\
X_3 &= (\xi_1 + \xi_2 - \xi_3)^2 \\
X_4 &= (\xi_1 + \xi_2 + \xi_3)^2
\end{aligned}
\qquad (31)
$$

Putting $4X_1' = X_1 - X_2 - X_3 + X_4$

$$
\begin{aligned}
4X_2' &= -X_1 + X_2 - X_3 + X_4 \\
4X_3' &= -X_1 - X_2 + X_3 + X_4 \\
4X_4' &= X_1 + X_2 + X_3 + X_4
\end{aligned}
\qquad (32)
$$

and then dropping the primes we get

$$
\begin{aligned}
X_1 &= 2\xi_2\xi_3 \\
X_2 &= 2\xi_1\xi_3 \\
X_3 &= 2\xi_1\xi_2 \\
X_4 &= \xi_1^2 + \xi_2^2 + \xi_3^2
\end{aligned}
\qquad (33)
$$

which is the same as equation 27.

We see then that the surface given by equation 11 which was an isoparametric basis function, that given by equation 18 which was a surface approximation implied by the isoparametric transformation method and the more general surface given by equations 28 are all the same surface which using a suitable coordinate system can be written as equation 33, or in any of the other parametric forms. This surface is called the Steiner surface after Jacob Steiner who first studied it in 1844 while on a journey to Rome. It is occasionally referred to as Steiner's Roman surface. A further discussion is given in [9] and the references therein contained. Before leaving this section we point out that all rational transformations imply that the images of planes are rational surfaces. In the polynomial case the Steiner surface is the simplest non-trivial one. Higher order transformations are in fact more common than the quadratic one which leads to the Steiner surface. Ninth order rational surfaces are then in common use though very little is known about them and almost nothing is known about the relevant approximation theory when using such surfaces as piecewise approximations to complicated geometry.

3. Max Noether's Theorem, Bezout's Theorem, And High Order Bases Instabilities.

Let us return to equation 7 which was the special case of equation 6 and gave a rational basis function corresponding to node 1 of figure 1a. This is the simplest example of a rational basis for a curved element. We will construct conforming basis functions. Though the word "conforming", has a more general meaning it will be sufficient for the purposes of this discussion to consider a basis as being a conforming one if each individual basis function is zero on all element sides which do not contain the corresponding node. From this definition it is easily deduced that conformity implies global C^0 continuity. We will also be constructing a basis which spans polynomials of some specified degree. Since the basis must span these polynomials globally i.e. over the entire element, for the basis has support of a single element, it must span polynomials around the element boundary. Thus we will seek a basis which maintains global C^0 continuity and which reduces to a polynomial of the required degree on the element sides.

We restrict ourselves to elements bounded by conic arcs. For a basis of order k we introduce $2k-1$ nodes between the vertices on each conic arc and an additional $(k-1)(k-2)/2$ interior nodes. Let the element have m conic arcs C_j $j = 1,\ldots m$ and n straight line segments L_j $j = 1,\ldots n$ with $m + n \geqslant 3$. Then if $N(m,n)$ is the number of nonvertex intersections in the plane produced from the entire set of intersections of the m conics and n lines, we have

$$N(m,n) = 4\sum_{i=1}^{m}(i-1) + 2mn + \sum_{i=1}^{n}(i-1) - (m+n)$$

$$= (2m+n-3)(2m+n)/2 \qquad (34)$$

Now an algebraic curve of degree n is uniquely determined by $n(n+3)/2$ points in general position and so we see that the nonvertex intersections of the sides of this element uniquely determine an algebraic curve of degree $2m + n - 3$. Let this curve have equation $D(x,y) = 0$ and refer to $D(x,y)$ as the denominator polynomial. Associated with a node i on one of the conic sides C_i say there is a unique algebraic curve passing through the remaining $2k - 2$ nodes on that conic and the $(k-1)(k-2)/2$ interior nodes for $2(k-1) + (k-1(k-2)/2 = \frac{(k-1)(k+2)}{2}$ which is the required number of points in general position to uniquely determine an algebraic curve or order $k - 1$. Let this curve have equation $A(x,y) = 0$ and refer to $A(x,y)$ as the "minimal adjacent polynomial." Also associated with this node let $O(x,y)$ be the product of all the polynomials associated with the equations of the element sides omitting only that polynomial associated with the conic containing node i. $O(x,y)$ is therefore of degree $2m + n - 2$. The algebraic curves and hence the associated polynomials are determined only up to a scalar factor. Let us assume that these factors have been chosen so that;

$$D(x_i,y_i) = A(x_i,y_i) = O(x_i,y_i) = 1 \tag{35}$$

Now define

$$W_i(x,y) = \frac{O(x,y)A(x,y)}{D(x,y)} \tag{36}$$

The rational function $W_i(x,y)$ given by equation 36 is suitable as a basis function associated with node i. Certainly,

$$W_i(x_j,y_j) = \delta_{ij}, \quad W_i(x,y) \equiv 0 \bmod C_k \quad k \neq i \tag{37}$$

$$W_i(x,y) \equiv 0 \bmod L_k \quad k$$

but it is also required that $W_i(x,y)$ reduce to a polynomial of degree k on the conic side C_i. The proof of this result is a simple application of Max Noether's theorem. The theorem requires certain conditions (Noether's conditions) to be satisfied concerning the intersections of three algebraic curves. We will not give detailed discussion of these conditions but refer the reader to texts on algebraic curves eg [4,18].

Theorem 1 (Max Noether). Let F, G and H be algebraic curves with F and G having no common components. Then there is an equation

$$H = AF + BG \tag{38}$$

with $\deg(A) = \deg(H) - \deg(F)$ and $\deg(B) = \deg(H) - \deg G$ if and only if Noether's conditions are satisfied.

We note here without proof that if the intersections of F and G and of H and G are simple and are the same set of intersections then Noether's conditions are satisfied. Bézout's theorem is also required. Again the reader is referred to texts on algebraic curves for more detailed discussion and proof.

Theorem 2 (Bezout). Algebraic curves of degrees m and n with no common components have mn intersections.

We must count the intersections with their appropriate multiplicities and we will make use of the case with multiple intersections in a later section.

We see therefore that if F and H have a common intersection set with G then they must be of the same degree in which case Noether's theorem gives

$$H = \alpha F + BG \tag{39}$$

where α is a scalar.

Now returning to the rational function $W_i(x,y)$ of equation 47 we note that $A(x,y)$ is of degree $k - 1$ and hence, by Bézout's theorem has $2k - 2$ intersections with C_i. By construction of $A(x,y)$ these intersections are the remaining $2k - 2$ nodes on C_i ie the nodes apart from node i. There are no other intersections of $A(x,y)$ with C_i. Also the set of intersections of $D(x,y)$ with C_i i.e. $D \cap C_i$ say, is included in the set of intersections of $O(x,y)$ with C_i i.e. $D \cap C_i \subset O \cap C_i$. There are precisely two additional intersection points in the set $O \cap C_i$ which are not in $D \cap C_i$ namely the two vertex points one at either end of the conic segment C_i. Let these two vertixes define a linear polynomial $L(x,y)$. Then from Noether's theorem we have

$$D(x,y)L(x,y) = \alpha O(x,y) + B(x,y)C_i(x,y) \tag{40}$$

where $C_i(x,y)$ is the polynomial associted with the conic C_i and α is a scaler. Hence when $C_i(x,y) = 0$ ie on the conic we have

$$D(x,y)L(x,y) \equiv O(x,y) \bmod C_i$$

Hence $$\frac{O(x,y)}{D(x,y)} \equiv L(x,y) \bmod C_i$$

Finally $$W_i(x,y) \equiv L(x,y)A(x,y) \bmod C_i \qquad (41)$$

which states the required result that $W_i(x,y)$ is equivalent to a polynomial of degree k on the conic segment. We note again that we know precisely the 2k intersections of $W_i(x,y) = 0$ with C_i viz the two vertices and the $2k - 2$ nodes other than node i. We illustrate this theory with a simple example. A more complete discussion is given in reference 17. In figure 4 let F, G and H be three conics and a, b, c the element concerned. Let us examine the case when $k = 3$ ie we seek a basis function which reduces to a cubic on F. We add 5 nodes on each conic segment and one interior node. There are nine external intersection points and D(x,y) is the unique cubic (up to a scalar factor) through these nine points. For node 1 say A(x,y) is the unique polynomial (again up to a scalar factor) determined by the five nodes 2, 3, 4, 5, and 6. $O(x,y) = G(x,y)H(x,y)$ where G(x,y) and H(x,y) are the polynomials of degree two associated with the conics G and H respectively.

Then

$$W_1(x,y) = \frac{G(x,y)H(x,y)}{D(x,y)} A(x,y) \qquad (42)$$

Now $$D \cap F = \{e_1, e_2, e_3, e_4, e_5, e_6\}$$

and $$GH \cap F = \{e_1, e_4, e_5, c, e_2, e_3, e_6, b\} \quad .$$

If we let the line bc have associated polynomial L(x,y) then we see immediately that $LD \cap F = GH \cap F$ and hence

$$W_1(x,y) \equiv L(x,y)A(x,y) \bmod F \qquad (43)$$

That is $W_1(x,y)$ does indeed reduce to a cubic polynomial on the conic F. The full discussion of the construction of the complete basis is given in the reference though we have given sufficient detail to appreciate the power of the geometry. We have also described the crucial quantities O(x,y), D(x,y) and A(x,y) in reasonable generality. The minimal adjacent polynomial is particularly relevant to our ensuing discussion.

It is reasonable to assume that when a particular type of basis is to be used for curved elements that it would be used for all such elements in the mesh. If we think of a two dimensional curved region we may think of having curved elements on the boundary. The actual curves will vary from element to element as we move around our domain. Some of the curves may be quite distinct and others may be very close to straight lines. We should therefore examine what happens to our basis as the curved side moves and finally flattens out to straight lines. The interpolating system is quite complicated as not only are the nodes moving but the actual space in which we are attempting our interpolation changes also. This situation is discussed in detail elsewhere [8] and we suffice with repeating one particularly significant result. We give a definition. If each basis function of a basis is well defined in the limit as the curved sides of an element reduce to straight lines we say that the basis is limit stable.

Theorem 3. If an element is bounded by straight lines and conics and if all the nodes remain distinct as the curve degenerates to a line then any basis of order k using the minimal adjacent polynomial can only be limit stable if $k = 1$.

Proof. For $k = 1$ it is shown in [8,16] that a limit stable basis exists. For $k > 1$. There are $2k - 1$ nodes between the vertices on each conic arc. The adjacent polynomial for node i say, is a polynomial of degree $k - 1$ which is zero at the remaining $2k - 2$ of these nodes and at an additional $(k-1)(k-2)/2$ interior nodes. Let one of the conics degenerate to a line. Then since the nodes remain distinct the adjacent polynomial of degree $k - 1$ now has $2k - 2$ zeros on the line. By Bezout's theorem this polynomial must be reducible with the linear form of the line as a factor and hence is also zero at node i. The corresponding basis function is then singular since it is now impossible to normalise the basis function to have unit value at that point. The basis is then not limit stable.

This theorem does not preclude the use of the minimal adjacent polynomial but it does show that one should not use it for elements whose curved sides are almost straight. In these cases one should either ignore the curve completely or use a limit stable basis which we develop later. We will also give an illustration of the above theorem but first we extend the application of the intersection theory.

4. A Corollary To Noether's Theorem, High Order Transformation Bases And A Limit Stable Basis.

The theorems of Bézout and Noether apply in the projective plane and also include complex intersections. This is necessary as we can see by the following simple example.

The equation

$$\alpha xy - x - y + 1 = 0 \tag{44}$$

represents a conic C say and the equation $y = 0$ represents a line L say By Bézout's theorem these two algebraic curves have two intersections but $y = 0 \Rightarrow x = 1$ which appears to lead to a single intersection at the point (1,0) ie

$$L \cap C = \{(1,0)\} \tag{45}$$

which appears to violate Bézout's theorem. However the curve given by equation 44 is only the affine representation of the corresponding algebraic curve whose equation is

$$\alpha xy - xw - yw + w^2 = 0 \tag{46}$$

written in the homogeneous coordinates of the plane x,y,w. Now $y = 0$ gives $w(x-1) = 0$ leding to the two intersections (1,0,1) and (1,0,0) ie

$$L \cap C = \{(1,0,1), (1,0,0)\} \tag{47}$$

and we have the expected number of intersections when we have included, where necessary, points of infinity. In a like manner we must include complex valued points. We notice then that intersections at infinity vanish when we restrict ourselves to the finite affine plane. This provides a corollary to Noether's theorem.

Corollary. Let F,G,H be algebraic curves satisfying Noether's conditions. Furthermore let $F \cap G = S$ where all the points in the intersection set are finite and let $F \cap H = S \cup S'$ where S' contains only points at infinity. Then $H \equiv G \bmod F|\text{aff}$, where the notation $F|\text{aff}$ is used to imply that we are considering equivalence in the finite affine plane.

This corollary together with the use of transformations provides a way of constructing a high order basis which has the advantage of a polynomial representation under the transformation.

In Figure 4 let the triangular element of 5a be transformed to that of 4b by the transformation of the first two of equations 13 ie by

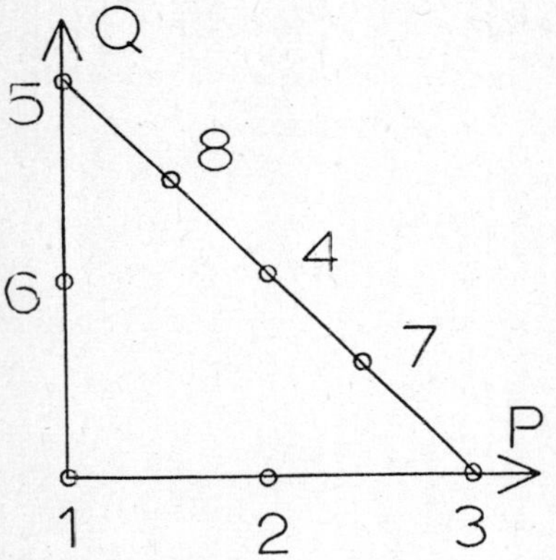

Figure 4a

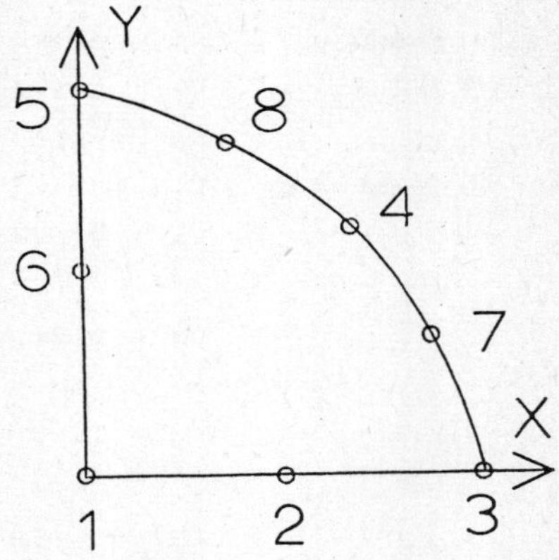

Figure 4b

$$x = p(1+\alpha q) \qquad y = q(1+\beta p) \tag{48}$$

where

$$\alpha = 2(2x_4-1) \quad \text{and} \quad \beta = 2(2y_4-1) \quad .$$

Let $C_i(x,y)$ $i = 1,2, \ldots, 6$ be polynomials of degree two defined by

$$C_i(x_j,y_j) = \delta_{ij} \quad i,j = 1,2, \ldots, 6 \tag{49}$$

Let $(A,B)_C$ denote the linear form defined by points A and B normalized to have unit value at C. (A,B and C are assumed noncolinear.) Let nodes 7 and 8 be defined as the images of points $(\frac{3}{4}, \frac{1}{4})$ and $(\frac{1}{4}, \frac{3}{4})$ under the transformation 48. Let

$$W_7(x,y,p,q) = \frac{16}{3}\, pq(8,4)_7$$

$$w_8(x,y,p,q) = \frac{16}{3}\, pq(7,4)_8 \tag{50}$$

$$w_i(x,y,p,q) = C_i(x,y) - C_i(x_7,y_7)\, w_7\,(p,q,x,y)$$

$$- C_i(x_8,y_8)\, w_8\,(p,q,x,y) \quad i = 1,2, \ldots, 6$$

This basis spans polynomials of degree two since the six $C_i(x,y)$ do. We wish to show that the basis is conforming. We require to show that W_7 and W_8 reduce to polynomials on the curved side in figure

5b. We apply Noether's theorem to do this. viz.
The curve 3,7,4,8,5 is the image under the transformation of the line $1-p-q = 0$ and is in fact a parabola. Now $p = 0$ has precisely one (and real) intersection with $1-p-q = 0$ at (0,1). Likewise $q = 0$ has precisely one intersection with $1-p-q = 0$ at (1,0). These intersection points are unaltered under the transformation. Let the parabola which is the image of $1-p-q = 0$ be denoted by f. Then

$$1-x-y \equiv p(1-p) \bmod f$$

but

$$pq \equiv p(1-p) \bmod f$$

$$pq \equiv (1-x-y) \bmod f$$

$$w_7(x,y,p,q) \equiv (1-x-y)(8,4)_7 \bmod f \tag{51}$$

Hence W_7 , and similarly W_8, reduce to quadratic polynomials on the curved side. In fact substitution shows that

$$1-x-y = -(\alpha+\beta)pq + 1-p-q$$

and hence

$$1-x-y = - (\alpha+\beta)pq \bmod f.$$

Of course $1-x-y$ is quadratic in p,q and has two finite intersections with the line $1-p-q = 0$. Such bases are called high order transformation bases [7].

We highlight the limit instability discussed in the last section and then use the corollary to Noether's theorem to produce a limit stable basis. As the curve degenerates to the line $1-x-y = 0$ the denominator polynomial $D(x,y)$ goes to unity and the adjacent polynomial reduces to $1-x-y$. For the usual rational basis

$$w_7 \to \frac{xy(1-x-y)}{x_7,y_7(1-x_7-y_7)} \to \infty$$

and similarly for the high order transformation basis above the limit is also unbounded. The difficulty, as pointed out in theorem 3, is the use of the minimal adjacent polynomial. A limit stable basis can be produced as follows. The image of $1-p-q = 0$ under the transformation given by equation 48 is a parabola , P , say whose axis has a slope β/α. Let L_8 be the line through node 8 with slope β/α and L_4 the line through node 4 with this same slope. Then by construction

$$(L_8 \; L_4) \cap f = \{8,4,\infty,\infty\} \tag{52}$$

where ∞ represents the point at infinity where either L_4 or L_8 meet the parabola again. By the corollary however

$$L_8 \; L_4 \equiv (8,4) \bmod f|aff \tag{53}$$

If we now choose

$$W_7^* (x,y,p,q) = L_8 \; L_4 \; p \; q \tag{54}$$

then $W_7^* \equiv W_7 \bmod f|aff$. However with the only restriction that as the curve degenerates to the line $1-x-y = 0$, $\beta/\alpha \to -1$ W_7^* will not become singular and we have a limit stable basis. We have thus made judicious use of Bézout's theorem, Noether's theorem and points at infinity to produce a limit stable high order basis. We end this section by noting that since x and y are polynomials in p and q we can write the entire basis as polynomials in p and q which may ease the integration problem.

5. Multiple Points and Hermite Interpolation

Much of what has been discussed in connection with the construction of Lagrange basis functions for curved elements can be generalized to produce Hermite basis. This generalization is given in the literature [6].

6. Parametric Curves as Piecewise Approximations

Parametrically defined curves and surfaces occur in numerical analysis in two very common ways. We have seen that in the finite element method most transformation techniques will implicitly use such curves and surfaces as approximations to curved regions. When transformations are being used it is not necessary to examine the nature of the geometric approximation in order to proceed with the finite element calculation. Although this facilitates the computation one has little idea of how good or bad is the implied approximation to the geometry . We have seen in the case of surfaces that the simplest nontrivial surface is the Steiner surface which has three double lines meeting in a single triple point. Another, and even more popular, occurrence of parametric curves and surfaces is in computer aided geometric design. The coordinates of the points having an explicit representation in terms of the parameter makes calculation simple. All such curves and surfaces of orders greater than two have singularities and hence, though easy to compute, may present difficulties in terms of providing good approximations or suitable design curves unless the positioning of the singularities can be controlled. We restrict ourselves for the remainder of this section to curves given parametrically in terms of rational functions and will further restrict most of our discussion to cubics whose parametrisations are polynomial. We start by indicating where such curves can arise.

If $y = S(x)$ is a spline curve interpolating data $\{y_i\}$ say then we can construct a parametric spline defined by

$$x = S(t), \quad y = S'(t) \tag{55}$$

where $S(t)$ interpolates data $\{x_i\}$ and $S'(t)$ interpolates data $\{y_i\}$. The parameter t could be chosen, for example, as chord length between the points $\{(x_i,y_i)\}$ or even as index i. Equation (70) then defines a rational curve, actually a polynomial curve since the parameterisations are polynomial. We use the terminology "polynomial curve" to denote a curve which permits a polynomial parameterisation. This includes curves which are polynomial functions in the usual sense but also includes nonfunction curves. The Bezier cubics, which we refer to again later, can be thought of as special cases of equation (55). The isoparametric transformation given by the first two numbers of equation (11) is of the form $x = X^*(p,q)$ $y = Y^*(p,q)$ and the image of the line $1-p-q = 0$ is of the form

$$x = X_2(p) \quad y = Y_2(p) \tag{56}$$

where $X_2(p)$ and $Y_2(p)$ are polynomials of degree two. The cubic counterpart of transformation 12 is given by

$$x = \sum_{i=1}^{10} x_i W_i(p,q) \qquad y = \sum_{i=1}^{10} y_i W_i(p,q) \tag{57}$$

where the $W_i(p,q)$ are cubic polynomials. Again we see that the image of any line is a polynomial cubic as is each piece of the usual parametric cubic spline. The Hermite version of equation (57) is interesting for it is implicit being of the form

$$x = \sum_{i=1}^{4} x_i W_i(p,q) + \sum_{i=1}^{3} \left(\frac{\partial x}{\partial p}\Big|_i U_i(p,q) + \frac{\partial x}{\partial q}\Big|_i V_i(p,q)\right)$$

$$y = \sum_{i=1}^{4} y_i W_i(p,q) + \sum_{i=1}^{3} \left(\frac{\partial y}{\partial p}\Big|_i U_i(p,q) + \frac{\partial y}{\partial q}\Big|_i V_i(p,q)\right) \tag{58}$$

where the set $\{W_i, U_i, V_i\}$ are the usual Hermite cubic basis for the triangle. Since in this case we must select the parameters $\frac{\partial x}{\partial p}\Big|_i$ etc. before we can use the transformation care must be exercised to choose these in such a way as to provide a satisfactory approximation to the curved boundaries. If a curve is given parametrically in the form

$$x = X_n(t,v) \quad y = Y_n(t,v) \cdot z = Z_n(t,v) \tag{59}$$

where $X_n(t,v)$, $Y_n(t,v)$, and $Z_n(t,v)$ are homogeneous polynomials of degree n in the homogeneous line coordinates t and v and x,y,z are the homogeneous coordinates in the plane then in general the curve is an n'th order rational curve. This can easily be seen since by Bézout's theorem the order of curve will be the same as the number of intersections between a line and the curve. Now a line is given by $ax + by + cz = 0$ and hence the intersections will be given by

$$aX_n(t,v) + bY_n(t,v) + cZ_n(t,v) = 0 \tag{60}$$

which, by the fundamental theorem of algebra will provide n ratios of t/v. Hence equation 56 is a conic and the curves implied by equations (57) and (58) are cubic curves. The conic given by equation (56) is in fact a parabola which we can see in the following way. Written in the form of equation (59), equation (56) becomes

$$\begin{aligned} x &= \alpha t^2 + \beta tv + \gamma v^2 \\ y &= \alpha' t^2 + \beta' tv + \gamma' v^2 \\ z &= v^2 \end{aligned} \tag{61}$$

The line $ax + by + cz = 0$ meets this conic when

$$(a\alpha+b\alpha')t^2 + (a\beta+b\beta')tv + (a\gamma+b\gamma'+c)v^2 = 0 \tag{62}$$

There is precisely one line given by $a/b = -\alpha'/\alpha$ such that the intersections are given by

$$v((a\beta+b\beta')t + (a\gamma+b\gamma'+c)v) = 0 \tag{63}$$

ie one intersection is finite and the other is at infinity. The conic must therefore be a parabola. Careful selection of the parameters which define the transformations, equation (57) or equation (58), could reduce the cubic curve to a conic or even restrict the transformation to a quadratic one in which case the implied curves will be parabolae. For example, consider equation (57) and figure 5. It was shown in [12] that choosing

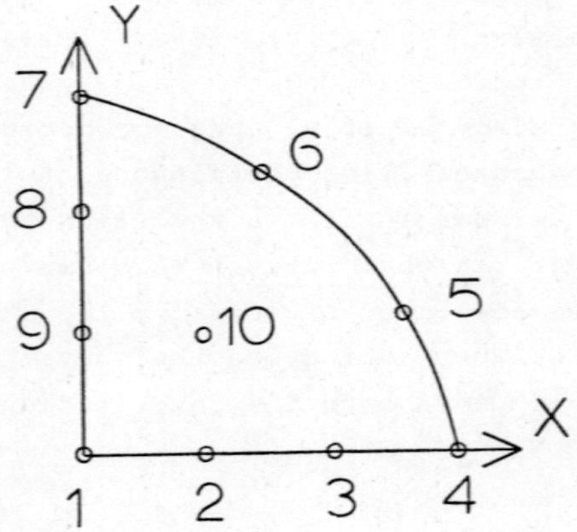

Figure 5.

$$\begin{aligned} x_5 &= x_6 + \tfrac{1}{3} \\ y_6 &= y_5 + \tfrac{1}{3} \end{aligned} \tag{64}$$

reduces the transformation equation (57) to

$$\begin{aligned} x &= p(1+\tfrac{9}{2}(6x_{10}-2x_6-\tfrac{4}{3})q + \tfrac{27}{2}\ (x_6+\tfrac{1}{3}-2x_{10})\ q\ (p+q)) \\ y &= q(1+\tfrac{9}{2}(6y_{10}-2y_5-\tfrac{4}{3})p + \tfrac{27}{2}(y_5+\tfrac{1}{3}-2y_{10})\ p\ (p+q)) \end{aligned} \tag{65}$$

which although still a cubic parametrisation the image of $1-p-q = 0$ is a parabola. A further restriction of

$$x_5 = 2x_{10}, \quad y_6 = 2y_{10} \tag{66}$$

reduces the transformation to

$$\begin{aligned} x &= p(1+3(3x_{10}-1)q) \\ y &= q(1+3(3y_{10}-1)p) \end{aligned} \tag{67}$$

The parabola has four degrees of freedom and can be specified by four linearly independent linear conditions. For example there is, in general, a unique parabola through four points and a unique parabola tangential to four given lines. In particular, we can define a parabola by two points and two lines passing through these points, that is, by Hermite interpolation at two points. This technique, if used in each interval between nodes on a closed curve, would yield a piecewise parabolic C' approximation. If $\{(x_i,y_i), f_i\}$ are the coordinates and corresponding slopes at nodes on a curve then in the interval between nodes i and $i + 1$ define a prabola by,

$$\alpha_i = 2\,\frac{(y_{i+1}-y_i) + (x_i-x_{i+1})(f_i+f_{i+1})}{f_i - f_{i+1}}$$

$$\beta_i = 2\,\frac{(x_i-x_{i+1})f_if_{i+1} - (y_i-y_{i+1})(f_i+f_{i+1}}{f_i - f_{i+1}}$$

$$x_i(t) = -\alpha_i t^2 + (\alpha_i+x(_{i+1}-x_i)t + x_i \tag{68}$$

$$y_i(t) = -\beta_i t^2 + (\beta_i+y_{i+1}-y_i)t + y_i$$

$$t\in[0,1]$$

This parabola is the Hermite interpolant at (x_i,y_i) and (x_{i+1},y_{i+1}) and hence a global C^1 curve is guaranteed. This technique can be used as a curve approximation method or as a way of selecting the required parameters in an isoparametric transformation in such a way as to ensure that the boundary approximation is C^1. For in equation (11) let us set

$$4x_4 = \alpha + 2(x_i+x_{i+1})$$
$$4y_4 = \beta + 2(y_i+y_{i+1}) \qquad (69)$$

and a C^1 boundary is guaranteed if this formula is applied in each element which contains a boundary segment. This method is described more fully elsewhere where error and a choice of norm is discussed, [12]. We note here that since parabolae (indeed conics in general) have no inflexion points it is advisable to select as nodes any inflexion points which occur in a curve which is being approximated by this method. We can avoid this restriction if we use higher order curves for then we can have inflexion points within the curves themselves. However using rational curves of degrees higher than two introduces another problem which is highlighted by the following theorem.

Theorem 4. The maximum number of double points of a non-degenerate curve of order n is $(n-1)(n-2)/2$ and if the curve has this number it is rational.

This indicates that in general rational curves possess multiple points. Indeed lines and conics are the only rational curves which do not have multiple points. It is therefore necessary that the positioning of these singularities be controlled for if care is not exercised a disasterously poor curve approximation may result. We will restrict the remainder of our discussion in this section to cubic curves.

In order to highlight the differences between the polynomial, the rational and the general algebraic cubics we mention first some of the relevant features of the general cubic curve and then specialise through the rational cubic to the polynomial one. It is the polynomial arc which is used in parametric cubic splines and in what have been called Bézier cubics. We have already seen that it is the polynomial cubic which is implied by any cubic transformation. For more information on cubic curves than we will give here the interested reader is referred to Primrose [14], Semple and Kneebone [16] or to any other standard text on algebraic curves.

Let $F(x,y,z)$ be an irreducible homogeneous polynomial of degree n in x,y,z with coefficients real or complex. The coordinates x,y,z form the homogeneous coordinates of a two dimensional projective space. By an irreducible algebraic curve we mean the set of all points whose coordinates (x,y,z) satisfy the equation

$$F(x,y,z) = 0 \qquad (70)$$

$F(x,y,z) = 0$ is called the equation of the curve and n is called the order of the curve. We can associate with a given algebraic curve a curve in the affine plane whose equation is given by

$$f(x,y) = F(x,y,1) = 0 \qquad (71)$$

which will give those points of the original curve which do not lie on the line at infinity. The locus of $f(x,y) = 0$ is referred to as the specialisation of $F(x,y,z) = 0$ to the affine plane or the affine representation of the curve. We refer to the "curve $F(x,y,z) = 0$" or "the curve F" rather than the "curve whose equation is $F(x,y,z) = 0$" and hope that this causes no confusion. Now from Bézout's theorem we see that a cubic cannot have a point of multiplicity greater than two for if it had then a line joining this point to any other point of the curve would have more than three intersections with the curve and would therefore be reducible. An inflexion point of a curve is a non-singular point of the curve at which the tangent line has at least three intersections with the curve. We exclude the degenerate case of a line. The Hessian $H(x,y,z) = 0$ of a curve F is a curve given by

$$H(x,y,z) = \begin{vmatrix} F_{xx} & F_{xy} & F_{xz} \\ F_{yx} & F_{yy} & F_{yz} \\ F_{zx} & F_{zy} & F_{zz} \end{vmatrix} = 0 \qquad (72)$$

We quote another theorem.

Theorem 5. The inflexions of a curve F are its non-singular intersections with its Hessian curve.

The Hessian of a cubic is another cubic and hence, a cubic has, in general, nine inflexions. Another theorem is important.

Theorem 6. If a rational curve has no points of multiplicity greater than two then it has its maximum number of double points.

We have seen that no cubic has a point of multiplicity greater than two and that the maximum number of double points of a cubic is one. Therefore a rational cubic has a double point and if a cubic has a double point it is rational. Hence all parametric cubic splines, Bézier cubics, and isoparametric cubics etc. have a double point.

There are two types of double points, nodes and cusps. A node is a double point at which there exists two distinct tangents. A double point where the tangents coincide is called a cusp. The inflexions of a rational curve $(X(t,v),\ Y(t,v),\ Z(t,v))$ are given by

$$\begin{vmatrix} X_{tt} & Y_{tt} & Z_{tt} \\ X_{tv} & Y_{tv} & Z_{tv} \\ X_{vv} & Y_{vv} & Z_{vv} \end{vmatrix} = 0 \tag{73}$$

and hence a rational curve of degree n has at most $3n - 6$ inflexions. In particular a rational cubic has at most three inflexions. Now it can be shown that the parameters of a cusp give a double root of equation (73) and hence a rational cubic with a cusp has but one inflexion. Now for a polynomial cubic

$$Z(t,v) = v^3 \tag{74}$$

and equation (73) is of the form

$$vQ_2(t,v) = 0 \tag{75}$$

where $Q_2(t,v)$ is a homogeneous polynomial of degree two. Four cases can occur:

i) $Q_2(t,v) = v(\alpha t+\beta v)$ $\alpha \neq 0$ in which case the curve has a cusp at infinity and an inflexion at $t/v = -\beta/\alpha$,

ii) $Q_2(t,v) = (\alpha t+\beta v)^2$ $\alpha \neq 0$ when there is a finite cusp and an inflexion at infinity,

iii) $Q_2(t,v) = \alpha t^2 + \beta tv + \alpha v^2$, $\beta^2 - 4\alpha\gamma > 0$ when there are two inflexions for real finite values of the ratio t/v and one inflexion at infinity,

iv) $Q_2(t,v) = \alpha t^2 + \beta tv + \gamma v^2$, $\beta^2 - 4\alpha\gamma < 0$ when two of the inflexions are not in the affine plane and the remaining one is at infinity.

It can be shown that when all the inflexions correspond to real values of the parameters then the node corresponds to a complex conjugate pair of ratios t/v and when the node corresponds to real values then two of the inflexions are in the complex plane. We can summarise the situation as follows. The non singular cubic has nine distinct inflexion points. A rational cubic has a cusp or a node. It has one inflexion if the double point is a cusp and three if the double point is a node. Restricting to the finite affine plane the polynomial cubic has either zero one or two inflexions, zero if the double point is nonisolated , one when it is at infinity (when it is a cusp) and two if it is isolated (when it is a node). This then induces four geometric equivalence classes of polynomial cubics which can be categorised as follows:

G1 Cusp at infinity

G2 Finite cusp in affine plane

G3 Finite non isolated node

G4 Isolated node

Representative members of each of these four cases are sketched in figure 6a, b, c, d respectively.

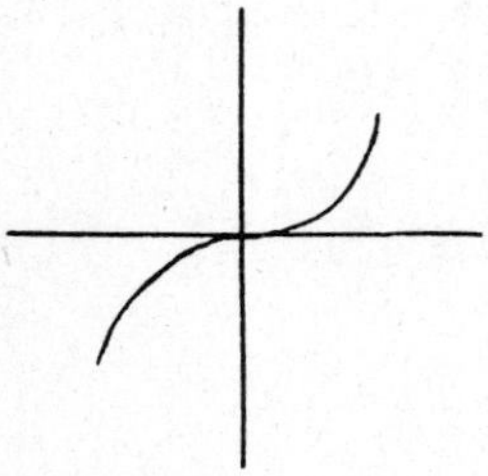

Figure 6a

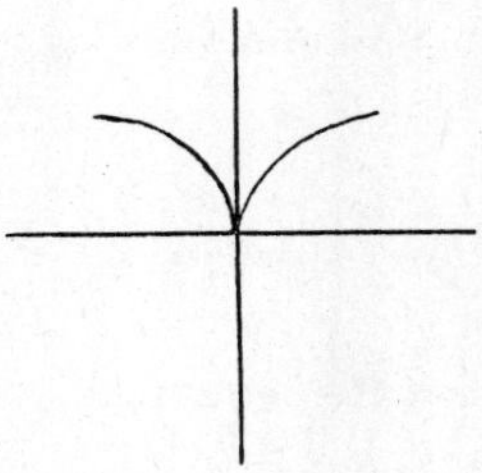

Figure 6b

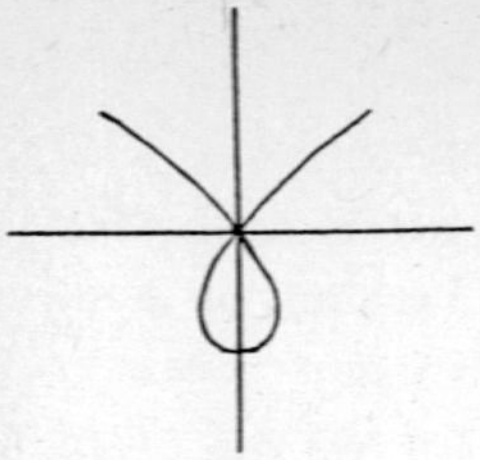

Figure 6c

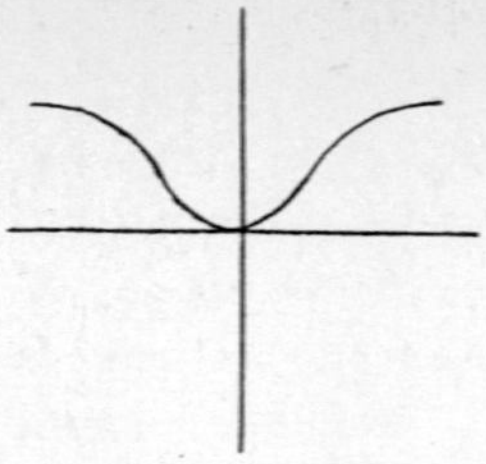

Figure 6d

We must now determine conditions on the coefficients which define the parameterisation which ensure that the curve is in the appropriate equivalence class. A partial classification of this problem has been given in the literature [3]. We outline here a more complete analysis.

Let us define the polynomial cubic in the Bézier form viz. Let $\vec{x}_0$, $\vec{x}_1$, $\vec{x}_2$, and $\vec{x}_3$ be the vector representations of four points in the plane. Then the Bézier form of the cubic is given by

$$\vec{P}(t) = \sum_{i=0}^{3} \binom{3}{i} (1-t)^{3-i} t^i \vec{x}_i \qquad t \in [0,1] \tag{76}$$

Let $\vec{\Delta} = \vec{x}_3 - \vec{x}_0$ and let $\vec{T}_0$ and $\vec{T}_1$ be the tangent directions at $\vec{x}_0$ and $\vec{x}_3$ respectively. We assume that $\vec{T}_0$ and $\vec{T}_1$ are normalised so that $\vec{T}_0 = \vec{\Delta} - \vec{T}_1$. Now define γ_0 and γ_1 by

$$\vec{x}_1 = \vec{x}_0 + \gamma_0 \vec{T}_0 \, , \quad \vec{x}_2 = \vec{x}_3 - \gamma_1 \vec{T}_1 \tag{77}$$

We can now rewrite equation (76) as

$$\vec{P}(t) = \vec{x}_0 + \vec{\Delta}\,(t^3+3t^2(1-t)+3\gamma_0 t(1-t)^2) - 3\vec{T}_1\,(\gamma_0 t(1-t)^2 + \gamma_1 t^2(1-t)) \tag{78}$$

For a node we have

$$\vec{P}(t_0) = \vec{P}(t_1) \qquad t_0 \neq t_1$$

giving

$$(t_1-t_0)\;Q\,(\vec{\Delta},\vec{T}_1,t_0,t_1) = 0$$

and hence

$$Q(\vec{\Delta},\vec{T}_1,t_0,t_1) = 0 \qquad (79)$$

Furthermore the conditions for a cusp give us

$$\underset{t_1 \to t_0}{\text{Limit}} \quad \frac{\vec{P}(t_1) - \vec{P}(t_0)}{t_1 - t_0} = Q(\vec{\Delta},\vec{T}_1,t_0,t_0) = 0$$

and hence putting $t_1 = t_0$ in equation (79) gives us the conditions for a cusp. Non degeneracy of the Bézier quadrilateral ensures that $\vec{\Delta}$ and $\vec{T}_1$ are not dependent and hence we deduce from equation (79) two equations which on solving for γ_0 and γ_1 give the conditions for a node in the range $t \in [0,1]$.

$$\gamma_0 = \frac{2t_1^2 + (2t_0-3)t_1 + 2t_0^2-3t_0}{3(t_1^2 + (t_0-2)t_1+(t_0-1)^2)}$$

$$(80)$$

$$\gamma_1 = \frac{2t_1^2 + (2t_0-3)t_1 + 2t_0^2 - 3t_0}{3(t_1^2+(t_0-1)t_1+t_0^2-t_0)}$$

$$t_0,t \in [0,1]$$

For a cusp in the range $t \in [0,1]$ we obtain

$$\gamma_0 = \frac{2t_0}{3t_0-1}, \quad \gamma_1 = \frac{2t_0-2}{3t_0-2} \qquad t_0 \in [0,1] \qquad (81)$$

Equation (80) represents a one parameter family of rational quartics in the γ_0, γ_1 plane given by

$$9t_0(t_0-1)\gamma_0^2\gamma_1^2 + 3t_0\gamma_0^2\gamma_1(3-2t_0) + 3\gamma_0\gamma_1^2(1-t_0-2t_0^2) + \gamma_0^2(1-t_0)^2$$

$$+ t_0^2\gamma_1^2 + \gamma_0\gamma_1(2t_0^2-2t_0-3) = 0 \qquad (82)$$

The region in the γ_0,γ_1 plane which corresponds to a node occurring in that part of the cubic which lies between $t = 0$ and $t = 1$ is given by $t_0,t_1 \in [0,1]$ in equation (80). The boundaries of this region correspond to $t_0 = 0$ and $t_0 = 1$ in which case the quartic is reducible giving

$$\gamma_0(3\gamma_1^2-3\gamma_1+\gamma_0) = 0$$

and (83)

$$\gamma_1(3\gamma_0^2-3\gamma_0+\gamma_1) = 0$$

respectively.

The condition for a cusp given by equation (81) represents part of the hyperbola

$$\gamma_0(3\gamma_1-4) - 4(\gamma_1-1) = 0 \tag{84}$$

These conditions give a complete characterisation of the γ_0,γ_1 plane into regions corresponding to no singularity, a cusp or a node occurring in the interval of interest ie for $t \in [0,1]$.

It is clear then that any method whatsoever which uses polynomial cubics as design curves or approximating curves must be such that the corresponding γ_0 and γ_1 lie in an appropriate region for the purposes in hand. The use of such curves together with the characterisation of the γ_0,γ_1 plane does however provide the opportunity of a local Hermite interpolation which can permit, unlike the parabolic case mentioned above, one or two inflexions between nodes. There will still be freedom to specify the γ_0 and γ_1 within the appropriate region and within that region each corresponding curve will have the desired singularity placement though will, of course, give different approximations to any given curve which is being interpolated. Some specific techniques for accomplishing such a local Hermite interpolation are discussed in more detail elsewhere [5].

7. A Geometric Approach to Blending Interpolants

We turn our attention to what has been called a blended interpolant or a transfinite blended interpolant. We will state the problem in the following way. Let R be a closed region in the x,y plane and let Γ be the boundary of R. Let a function f be given on Γ though not in R. We assume that f is continuous though this condition could be relaxed. Find a function $B(f)$ which is defined on R and equals f on Γ ie

$$\dot{B}(f) = f \bmod \Gamma \tag{85}$$

This problem has received much attention in recent years with attempts being made to extend the original approach to more general regions R. A survey and bibliography is given in [2]. The usual approach is to construct certain "projectors" and then to take a Boolean sum. This procedure works well for simple regions, particularly product regions, but does not readily extend to more general geometries. Here we take a different approach to the construction of the interpolant. In the simpler cases this method will yield identical results to the traditional method however it can produce blended interpolants in cases where suitable projectors necessary in the Boolean sum approach are not known. This technique is discussed in more detail elsewhere [9] and we satisfy ourselves here by outlining the salient features of the geometric application.

We start by constructing a planar extension of f. The function f is defined only on Γ and we seek to define a function of the two space variables. Let R comprise segments of N curves whose equations are

$$\gamma_i(x,y) = 0 \quad i = 1, 2, \ldots, N \tag{86}$$

We assume that these are algebraic curves though this restriction is not strictly necessary. Γ is then the subset of the $\bigcup_i \gamma_i$ which contains the appropriate segments. Let f_i denote that part of f which is defined on the boundary segment of γ_i. The vertices of R whose coordinates we label $\{(x_i,y_i)\}$ $i = 1,2,\ldots,N$ lie on Γ and are assumed to satisfy

$$\begin{aligned} (x_1,y_1) &\in \gamma_N \cap \gamma_1 \\ (x_i,y_i) &\in \gamma_{i-1} \cap \gamma_i \quad i = 2, 3, \ldots, N \end{aligned} \tag{87}$$

Now use the symbol μ_i for $\gamma_i(x,y)$. Then the equations

$$\begin{aligned} \gamma_{i-1}(x,y) - \mu_{i-1} &= 0 \\ \gamma_i(x,y) - \mu_i &= 0 \end{aligned} \tag{88}$$

are polynomial in x and y and by straightforward elimination theory, ie Sylvester's dialytic elimination, we can eliminate y or x from these two polynomials to obtain

$$\hat{X}(\mu_{i-1}, \mu_i, x) = 0$$

and (89)

$$\hat{Y}(\mu_{i-1}, \mu_i, y) = 0$$

respectively.

Now the intersection cycle of γ_{i-1} and γ_i is the set of common zeros of the equation $\gamma_{i-1}(x,y) = \gamma_i(x,y) = 0$ and hence can be obtained by elimination of y or x from these two polynomials yielding $\hat{X}(0,0,x) = 0$ and $\hat{Y}(0,0,y) = 0$ respectively. By definition the vertex (x_i,y_i) belongs to this intersection cycle. Hence $(x-x_i)$ is a factor of $\hat{X}(0,0,x)$ and $(y-y_i)$ factor of $\hat{Y}(0,0,y)$. We can therefore solve equations (89) for x and y to obtain

$$x = X(\mu_{i-1}, \mu_i) \quad \text{with} \quad x_i = X(0,0)$$

(90)

$$y = Y(\mu_{i-1}, \mu_i) \quad \text{with} \quad y_i = Y(0,0)$$

or we could arrange for $X(0,0) = x_{i-1}$, $Y(0,0) = y_{i-1}$.

If f_i is given as $f_i(x,y)$, though defined only on a segment of γ_i, we can make planar extensions by setting

$$f_{i,i-1} = f_i(X(\mu_{i-1},0), \ Y(\mu_{i-1},0))$$

(91)

$$f_{i-1,i} = f_{i-1}(X(0,\mu_i), \ Y(0,\mu_i))$$

We notice that both $f_{i,i-1}$ and $f_{i-1,i}$ are defined in the plane and that

$$f_{i,i-1} = f_i \bmod \gamma_i$$

and (92)

$$f_{i-1,i} = f_{i-1} \bmod \gamma_{i-1} .$$

In addition $\gamma_i = 0 \Rightarrow \mu_i = 0$

hence $f_{i-1,i} = f_i(X(0,0), \ Y(0,0)) \bmod \gamma_i$

$$= f_i(x_i,y_i) \bmod \gamma_i \quad j = i \text{ or } i-1 .$$

That is $$f_{i-1,i} = f(x_i, y_i) \bmod \gamma_i . \qquad (93)$$

We read $f_{i,i-1}$ as "f_i on i-1."

Since both nodes i and $i-1$ lie on γ_{i-1} and we may wish to use equation 90 with $X(0,0) = x_i$ or with $X(0,0) = x_{i-1}$ we can introduce the notation $f^j_{i-1,i}$ to indicate that $X(0,0) = x_j$ and $Y(0,0) = y_j$. This removes the confusion. Usually $j = i$ in which case we drop the superscript ie $f_{i\pm1,i}$ and $f_{i,i\pm1}$ are taken to mean $f^i_{i\pm1,i}$ and $f^i_{i,i\pm1}$ respectively.

In the cases where the curves γ_i and functions f_i are given only as data we can construct an interpolating or approximating curve say a parametric spline or piecewise parabolic approximation to Γ and an interpolant to each f_i. It may be possible, and easy to check, that the approximating γ_{i-1} and γ_i could be used as independent variables to construct an interpolant to the f_i. That is we may be able to construct the $f_{i-1,i}$ and $f_{i,i-1}$ directly. A very likely occurrence is the situation where we have a transformation say

$$x = \bar{X}(p,q) \qquad y = \bar{Y}(p,q) \qquad (94)$$

under which each γ_i is the image of a line in the p,q plane. In this case each μ_i is linear in p and q. Hence we have from equation (88)

$$\mu_{i-1} = \alpha_{i-1}p + \beta_{i-1}q + \gamma_{i-1}$$

$$\mu_i = \alpha_i p + \beta_i q + \gamma_i$$

Solving for p and q we get

$$p = a_{i-1}\mu_{i-1} + b_{i-1}\mu_i + c_{i-1}$$

$$q = a_i\mu_{i-1} + b_i\mu_i + c_i$$

giving from equation (94)

$$x = \bar{X}(p,q) = X(\mu_{i-1},\mu_i)$$

$$y = \bar{Y}(p,q) = Y(\mu_{i-1},\mu_i)$$

and the planar extensions are hence most easily defined.

We now take a basis $\{\phi_j(x,y)\}$ $j = 1,2, \ldots, n \geq N$ and additional points $\{(x_i,y_i)\}$ $i = N+1, \ldots, n$ which satisfies the following properties:

i) $$\phi_j(x_i,y_i) = \delta_{ij} \quad i,j = 1, 2, \ldots, n$$

ii) $$\phi_j(x,y) \equiv 0 \bmod \gamma_i \quad j = 1, 2, \ldots, N \quad i \neq j-1 \text{ or } j$$

$$\phi_j(x,y) \equiv 0 \bmod \gamma_i \quad j = N+1, N+2, \ldots, n \quad i = 1,2, \ldots, N$$

$$\phi_j(x,y) \equiv 0 \bmod \gamma_i \quad i = 1,2, \ldots, N \quad \text{iff} \quad (x_j,y_j) \in \Gamma \tag{95}$$

iii) $$\sum_{j=1}^{n} \phi_j(x,y) = 1 \bmod \Gamma$$

Each $\phi_j(x,y)$ is then associated with a point (x_j,y_j) which we refer to as the node (x_j,y_j) corresponding to $\phi_j(x,y)$ or simply as node j. For the vertex nodes $1, 2, \ldots, N$ the corresponding $\phi_j(x,y)$ are non zero on the adjacent sides γ_{j-1} and γ_j and zero on all other boundary segments. For non vertex nodes on a boundary segment the corresponding $\phi_j(x,y)$ is non zero on that segment but zero on all other segments. For nodes not on Γ the corresponding $\phi_j(x,y)$ are zero on Γ. We simplify the notation by writing ϕ_j for $\phi_j(x,y)$ and by introducing the following sets. For each $i = 1, 2, \ldots, N$ let C_i denote the set of indices corresponding to nodes on γ_i. We note that this will certainly include the two vertex nodes i and $i+1$ (modulo N) and may include other indices greater than N. Let D_i denote the set of these other indices and let E_Γ denote the set of indices corresponding to nodes not on Γ. Now define $B(f)$ by

$$B(f) = \sum_{i=1}^{N} \left| f_i \sum_{j\in D_i} \phi_j + \phi_i \; (f_{i,i-1}+f_{i-1,i}-f(x_i,y_i)) \right| + \sum_{i\in E_\Gamma} \alpha_i\phi_i \tag{96}$$

<u>Theorem 7.</u> $B(f)$ is an interpolant to f on the boundary Γ of R.

Proof. Consider $B(f) \bmod \gamma_k$. From equation (95) we have for $j \in D_i$ $\phi_j = 0 \bmod \gamma_k$ unless $k = i$. Hence

$$B(f) \equiv \left| f_k \sum_{j \in D_k} \phi_j + \phi_k (f^k_{k,k-1} + f^k_{k-1,k} - f(x_k, y_k)) \right.$$

$$\left. + \phi_{k+1} (f^{k+1}_{k+1,k} + f^{k+1}_{k,k+1} - f(x_{k+1}, y_{k+1})) \right| \bmod \gamma_k$$

Now from the properties of the planar extensions equations (95) and (93) we have

$$f^k_{k,k-1} = f^{k+1}_{k,k+1} = f_k \bmod \gamma_k$$

$$f^k_{k-1,k} = f(x_k, y_k) \bmod \gamma_k$$

and

$$f^{k+1}_{k+1,k} = f(x_{k+1}, y_{k+1}) \bmod \gamma_k \;.$$

Hence

$$B(f) = \left| f_k \sum_{j \in D_k} \phi_j + \phi_k f_k + \phi_{k+1} f_k \right| \bmod \gamma_k$$

$$= \left| f_k \sum_{j \in C_k} \phi_j \right| \bmod \gamma_k \;.$$

Using again the properties of the $\{\phi_j\}$, equation (95) since

$$\sum_{j=1}^{n} \phi_j = 1 \bmod \Gamma$$

then

$$\sum_{j \in C_k} \phi_j = 1 \bmod \gamma_k$$

Hence

$$B(f) = f_k \bmod \gamma_k.$$

Since this is true for each $k = 1, 2, \ldots, N$ we have finally

$$B(f) = f \bmod \Gamma \;.$$

Special cases and further simplifications are given in the references cited. We give one example here.

Let R be the right hand semi-circle bounded by $\gamma_1 \equiv x = 0$ and $\gamma_2 \equiv x^2 + y^2 - 1 = 0$ figure 7. Choose a basis of three

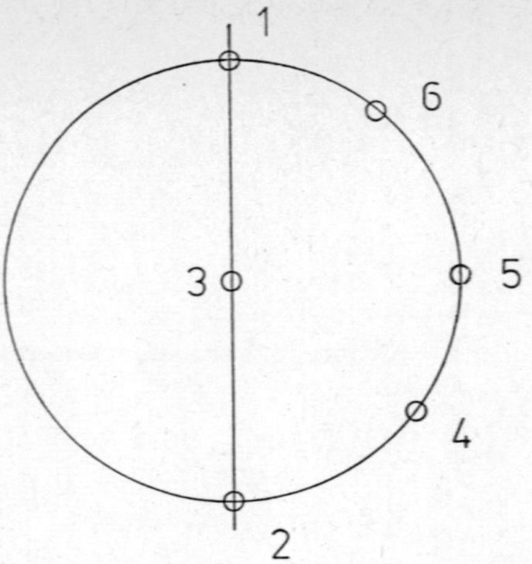

Figure 7.

functions given by

$$\phi_1 = \frac{1}{2}(1-x-y)$$

$$\phi_2 = \frac{1}{2}(1-x-y) \qquad (97)$$

$$\phi_5 = x \quad .$$

Following equation (88) we set

$$\mu_i - x = 0 \quad \text{and} \quad \mu_2 - x^2 + y^2 + 1 = 0$$

giving, see equation (90), $x = \mu_1$ and $y = \pm(1+\mu_2-\mu_1^2)^{1/2}$; we take the positive sign for that term in $B(f)$ which is multiplied by ϕ_1 and the negative sign for the term which is multiplied by ϕ_2. That is we are choosing the $X(\mu_{i-1},\mu_i)$ $Y(\mu_{i-1},\mu_i)$ of equation (90) to satisfy $(X(0,0),\ Y(0,0)) = (0,\ \pm1)$ as suits our needs. Then

$$f_{12}^1 = f(0,(1+\mu_2)^{1/2}), \quad f_{21}^1 = f(\mu_1,(1-\mu_1^2)^{1/2})$$

$$f_{12}^2 = f(0,-(1+\mu_2)^{1/2}), \quad f_{21}^2 = f(\mu_1,-(1-\mu_1^2)^{1/2}) .$$

These functions satisfy conditions 92 and 93.
In this case D_1 is empty and $D_2 = \{5\}$ and $B(f)$ is given by

$$B(f) = f_2\phi_5 + \phi_1[f(0,(x^2+y^2)^{1/2}) + f(x,(1-x^2)^{1/2}) - f(0,1)]$$

$$+ \phi_2[f(x,-(1-x^2)^{1/2}) + f(0,-(x^2+y^2)^{1/2}) - f(0,-1)]$$

We could use any basis which satisfies conditions 95. For example choose the six basis functions given by:

$$\phi_1 = (2;\ 3;\ 4;\ 5;\ 6)_1$$

$$\phi_2 = (1;\ 3;\ 4;\ 5;\ 6)_2$$

$$\phi_3 = (1;\ 2;\ 4;\ 5;\ 6)_3$$

$$\phi_4 = (1;\ 2)_4(5;\ 6)_4$$

$$\phi_5 = (1;\ 2)_5(4;\ 6)_5$$

$$\phi_6 = (1;\ 2)_6(4;\ 5)_6$$

This set satisfies the required properties as can be shown by applying Bézout's theorem. Here $D_1 = \{3\}$ and $D_2 = \{4,5,6\}$ and another suitable interpolant is given by

$$B(f) = f_1\phi_3 + f_2(\phi_4+\phi_5+\phi_6) + f_1(f_{12}^1 + f_{21}^1 - f(0,1))$$

$$+ f_2(f_{21}^2 + f_{12}^2 - f(0,-1) \quad .$$

REFERENCES

1. Baker, H. F., An Introduction to Plane Geometry. Originally published by Cambridge University Press. Published again (1971) by Chelsea Publishing Company.

2. Barnhill, R. E., Blending function interpolation: a survey and some new results, University of Dundee, Department of Mathematics, Numerical Analysis Report, No. 9, July 1975.

3. Ferguson, J. C., Multivariate Curve Interpolation, J. ACM 11, 2, 221-228 (1964).

4. Fulton, W., Algebraic Curves: An Introduction to Algebraic Geometry, W. A. Benjamin, Inc. 1969.

5. Hickman, M. G., and McLeod, R. J. Y., Polynomials Cubics and Piecewise Approximation of Curves. To be published.

6. McLeod, R., Hermite Interpolation Over Curved Finite Elements, J. Approx. Theory, V. 19, No. 2, Feb. 1977, 101-117.

7. McLeod, R., High Order Transformation Methods for Curved Finite Elements, J. Inst. Math. Applics. (1978) 21, 419-428.

8. McLeod, R. J. Y., On the stability of two-dimensional interpolation and high order bases for curved finite elements, Conf. Math. Applics. V. 5, No. 4, 1979, 249-266.

9. McLeod, R. J. Y., The Steiner Surface Revisited, Proc. R. Soc. Lond. A 369 157-174 (1979).

10. McLeod, R. J. Y., A Geometric Approach to Blended Interpolants. To be published.

11. McLeod, R. and Mitchell, A. R., The Construction of Basis Functions for Curved Elements in the Finite Element Method. JIMA (1972) 10, 382-393.

12. McLeod, R. J. Y., and Mitchell, A. R., The use of parabolic arcs in Mateling curved boundaries in the finite element method. J. Inst. Math. Applics. (1975) 16, 239-246.

13. McLeod, R. J. Y., and Mitchell, A. R., A piecewise parabolic C' approximation for curved boundaries, Conf. Math. Applics., V. 5, No. 4 (1979) p. 277-284.

14. Primrose, E. J. F., Plane Algebraic Curves, MacMillan and Co. Ltd., London, 1955.

15. Salmon, G., A Treatise on Conic Sections. There are many printings of this text. A recent 6th edition has been published by Chelsea Publishing Company, New York, N. Y.

16. Semple, J. G. and Kneebone, G. T., Algebraic Curves, Oxford, University Press, 1959.

17. Wachspress, E. L., A Rational Finite Element Basis, Academic Press (1975).

18. Walker, R. J., Algebraic Curves, Dover Publications Ltd., New York, 1962.(Originally published by Princeton University Press in 1950.)

A QUASIOPTIMAL ESTIMATE IN PIECEWISE POLYNOMIAL GALERKIN APPROXIMATION OF PARABOLIC PROBLEMS

Lars B. Wahlbin

1. INTRODUCTION.

For the purposes of this exposition we shall consider only semi-discrete spline-Galerkin approximations of a simple model problem. Thus let $u(t,x)$, $t \geq 0$, $0 \leq x \leq 1$, solve the problem

(1.a) $\quad u_t = u'' + f(t,x), \quad t \geq 0, \quad 0 \leq x \leq 1,$

(1.b) $\quad u(t,0) = u(t,1) = 0, \quad t \geq 0,$

(1.c) $\quad u(0,x) = v(x), \quad 0 \leq x \leq 1 .$

Here $f(t,x)$ and $v(x)$ are given functions, and we use the notation $u_t = \partial u/\partial t$, $u' = \partial u/\partial x$, $u'' = \partial^2 u/\partial x^2$. When $f(t,x) \equiv 0$ in (1.a) we shall write

$$u(t,x) = E(t)v(x) ,$$

thus defining the homogeneous solution operator $E(t)$.

For the numerical solution of (1) we again consider only a simple case. For N an integer, let

$$h = 1/N; \; x_i = ih, \quad i = 0,\dots,N; \quad I_i = [x_i, x_{i+1}], \quad i = 0,\dots,N-1.$$

For $r \geq 2$ an integer, set

$$S_h = \{\chi \in \mathcal{C}^0[0,1], \chi(0) = \chi(1) = 0, \; \chi\big|_{I_i} \text{ is a polynomial of degree } r-1\} .$$

As a continuous in time approximation to the solution of (1), let $u_h(t):[0,\infty) \to S_h$ be given by

(2.a) $\quad ((u_h)_t, \chi) + (u_h', \chi') = (f,\chi), \text{ for } \chi \in S_h ,$

(2.b) $\quad u_h(0) = v_h \in S_h ,$

where $(u,v) = \int_0^1 u(x)\overline{v(x)}dx$ and v_h is a given approximation to initial data v. If $f \equiv 0$ in (2.a) we have the discrete solution operator by

$$u_h(t,x) = E_h(t)v_h(x) .$$

We shall prove the following "quasioptimal" error estimate.

THEOREM.

There exist constants K^1 and K^2 such that for $h \le 1/2$, $t > 0$,

$$(3) \qquad \max_{0 \le s \le t} \|u_h(s) - u(s)\|_{L_\infty} \le K^1 \ln 1/h \, \|v_h - v\|_{L_\infty} + K^2 \ln 1/h(1 + \ln(1 + t/h^2)) \max_{0 \le s \le t} \min_{\chi \in S_h} \|\chi - u(s)\|_{L_\infty} .$$

Remark 1.

The constants in (3) are probably not best possible. From the proof other information can be obtained, e.g., for t large, influence of the initial approximation is weighted down. With $k > 0$, K^1, K^2 and K^3 constants one has for $t \ge 1$,

$$\|u_h(t) - u(t)\|_{L_\infty} \le K^1 e^{-kt} \|v_h - v\|_{L_\infty} + K^2 e^{-kt} \max_{0 \le s \le t/2} \min_{\chi \in S_h} \|\chi - u(s)\|_{L_\infty}$$

$$+ K^3 \ln 1/h(1 + \ln(1 + t/h^2)) \max_{t/2 \le s \le t} \min_{\chi \in S_h} \|\chi - u(s)\|_{L_\infty} ;$$

this serves as an example.

Following [3], let us outline some stepping stones towards an estimate like (3). With $\|\cdot\|_B$ a Banach space norm we seek an estimate

$$(\mathrm{Q}) \qquad \max_{0 \le s \le t} \|u_h(s) - u(s)\|_B \le C_Q^1 \|v_h - v\|_B + C_Q^2 \max_{0 \le s \le t} \min_{\chi \in S_h} \|\chi - u(s)\|_B .$$

Now C_Q^i, $i = 1,2$, could depend on t and h, but, one hopes, only in a mild way.

The method for showing (Q) proceeds via the following four estimates, (S), (A), (P) and (Π) below. The first building block is a stability estimate

$$(\mathrm{S}) \qquad \|E_h(t) v_h\|_B \le C_S \|v_h\|_B, \quad t \ge 0 .$$

The second estimate is essentially the fact that the semigroups $E_h(t)$ are analytic "uniformly" in h,

$$(\mathrm{A}) \qquad \left\|\frac{\partial}{\partial t} E_h(t) v_h\right\|_B \le \frac{C_A}{t + h^2} \|v_h\|_B, \quad t \ge 0 .$$

The two remaining estimates pertain to the generators of the semigroups $E(t)$ and $E_h(t)$ and a connection between them. The generator of $E(t)$ is $D^2 = \partial^2/\partial x^2$ with suitable domain. The discrete counterpart on S_h is $D_h^2 \colon S_h \to S_h$ given by

$$(4) \qquad (D_h^2 f_h, \chi) = -(f_h', \chi'), \text{ for } \chi \in S_h ,$$

and D_h^2 is the generator of $E_h(t)$. The two generators are related by use of two projection operators,

$$(5) \qquad D_h^2 P_h f = \Pi_h D^2 f$$

where $P_h : L_\infty \to S_h$ is the H_0^1 projection given by

$$((P_h f)', \chi') = (f', \chi')$$

$$= \sum_{i=0}^{N-1} \left(- \int_{I_i} f\chi'' dx + f\chi' \Big|_{x_i}^{x_{i+1}}\right), \text{ for } \chi \in S_h,$$

and $\Pi_h : L_1 \to S_h$ is the L_2 projection,

$$(\Pi_h f, \chi) = (f, \chi), \text{ for } \chi \in S_h .$$

It is easy to verify (5). For,

$$(D_h^2 P_h f, \chi) = -((P_h f)', \chi') = -(f', \chi') = (f'', \chi) = (\Pi_h f'', \chi), \text{ for } \chi \in S_h.$$

We assume that we know the following two stability type results for the projections.

$$(P) \qquad \|P_h f\|_B \le C_P \|f\|_B ,$$

and

$$(\Pi) \qquad \|\Pi_h f\|_B \le C_\Pi \|f\|_B .$$

Let us show how (Q) follows from (S), (A), (P) and (Π). The semidiscrete Galerkin solution of (2) is given by

$$(6) \qquad \begin{aligned} &(u_h)_t - D_h^2 u_h = \Pi_h f, \quad t \ge 0, \\ &u_h(0) = v_h . \end{aligned}$$

Write the error as

$$(7) \qquad u_h(t) - u(t) = u_h(t) - P_h u(t) + (P_h - I) u(t) .$$

Here by (P),

$$(8) \qquad \|(P_h - I) u(t)\|_B = \|(P_h - I)(u(t) - \chi)\|_B \le (C_P + 1) \|u(t) - \chi\|_B, \text{ for any } \chi \in S_h.$$

Let $\theta(t) = u_h(t) - P_h u(t)$; as an equation in S_h we have by use of (6), (5) and (1),

$$\theta_t - D_h^2\theta = (u_h)_t - D_h^2 u_h - (P_h u)_t + D_h^2 P_h u = \Pi_h f - (P_h u)_t + \Pi_h u''$$

$$= \Pi_h f + ((\Pi_h - P_h)u)_t + \Pi_h(u'' - u_t) = ((\Pi_h - P_h)u)_t .$$

Therefore by Duhamel's principle and by integration by parts,

(9)
$$\theta(t) = E_h(t)(v_h - P_h v) + \int_0^t E_h(t-s)\frac{\partial}{\partial s}(\Pi_h - P_h)u(s)ds$$
$$= E_h(t)(v_h - P_h v) + (\Pi_h - P_h)u(t) - E_h(t)(\Pi_h - P_h)v + \int_0^t \frac{\partial}{\partial s} E_h(t-s)(\Pi_h - P_h)u(s)ds .$$

The terms involving v_h and v combine to $E_h(t)\Pi_h(v_h - v)$ and so by (S) and (Π),

(10)
$$\|E_h(t)\Pi_h(v_h - v)\|_B \leq C_S C_\Pi \|v_h - v\|_B .$$

Also, for any $\chi \in S_h$,

$$(\Pi_h - P_h)u(t) = (\Pi_h - P_h)(u(t) - \chi)$$

so that

(11)
$$\|(\Pi_h - P_h)u(t)\|_B \leq (C_\Pi + C_P)\|u(t) - \chi\|_B, \text{ any } \chi \in S_h .$$

Finally, by (A), (P) and (Π) for any $\chi(s)$ in S_h,

$$\left\|\int_0^t \frac{\partial}{\partial s} E_h(t-s)(\Pi_h - P_h)u(s)ds\right\|_B \leq \int_0^t \frac{C_A}{t-s+h^2}(C_\Pi + C_P)\|u(s) - \chi(s)\|_B ds$$

$$\leq C_A(C_\Pi + C_P)\ln(1 + t/h^2) \max_{0 \leq s \leq t} \min_{\chi \in S_h} \|u(s) - \chi\|_B .$$

Combining this with (9), (10) and (11) and reporting the result of that, together with (8) into (7), we obtain (Q) with

(12)
$$C_Q^1 = C_S C_\Pi$$
$$C_Q^2 = (C_P + 1) + (C_P + C_\Pi)(1 + C_A \ln(1 + t/h^2)) .$$

As for Banach spaces B where one might obtain (S), (A), (P) and (Π), we remark that it is fairly easy for $B = H^1$. For $B = L_2$, the estimate (P) fails, unless we work with smoother spaces S_h.

Our Theorem amounts to (Q) with $B = L_\infty$ and certain constants. It is well known, cf. [1], [2], [6], that in the one-dimensional maximum norm case C_P and C_Π are absolute constants, independent of h. We shall prove in the second part of this paper that

PROPOSITION.

For $B = L_\infty(0,1)$ there are absolute constants K_S and K_A such that for $0 < h \leq 1/2$, (S) and (A) hold with

$$C_S = K_S \ln 1/h,$$
$$C_A = K_A \ln 1/h .$$

Clearly, by what we have said above the Theorem would follow from this Proposition.

Our method of proof is to obtain the bounds of the Proposition from a careful estimate of the discrete resolvent. In [3], in a case of two space dimensions and $B = L_\infty$, the Proposition was proved in a more direct way; in fact, a resolvent estimate was derived as a by-product of (S) and (A).

The discrete resolvent estimate, (R) of the second part, will be derived by methods inspired by those of [4]. It would be very easy, see Remark 3 in particular, to treat a singular perturbation problem of "elliptic to elliptic type", e.g., $-\varepsilon u''+u = f$, in a similar way. In [5], the Theorem will be extended to certain problems with time dependent coefficients.

We remark that the estimate (A) is useful in connection with numerical use of the smoothing property in parabolic problems, and that the resolvent estimate (R) could be used to investigate certain time discretizations.

2. PROOF OF THE PROPOSITION.

We assume that the reader is conversant with inverse properties and approximation results for the spaces S_h.

To prove the Proposition, i.e., (S) and (A) with appropriate constants, we shall use a resolvent estimate. With

$$R_h^z = (z - D_h^2)^{-1} ,$$

cf. (4), we have

$$E_h(t)v_h = \frac{1}{2\pi i} \int_{\Gamma_h} e^{tz} R_h^z v_h \, dz ,$$

$$\frac{\partial}{\partial t} E_h(t)v_h = \frac{1}{2\pi i} \int_{\Gamma_h} z e^{tz} R_h^z v_h \, dz .$$

Here, with c_1 and C_1 suitable positive constants, $\Gamma_h = \Gamma_h^1 \cup \Gamma_h^2 \cup \Gamma_h^3 \cup \Gamma_h^4$ is the curve in the complex plane indicated below.

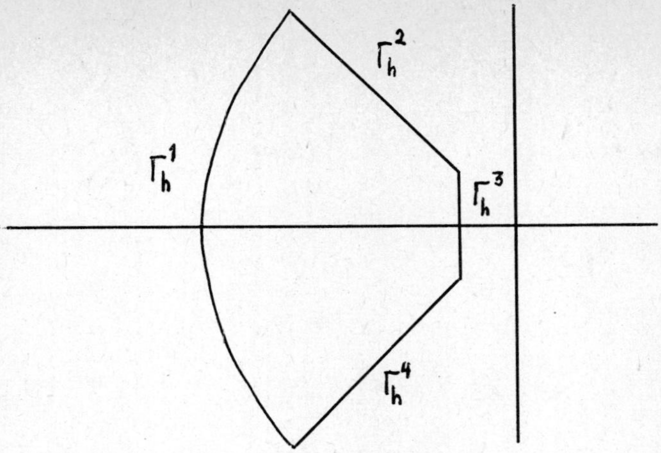

Γ_h^1: $|z| = C_1 h^{-2}$,

Γ_h^2: Re z = -Im z,

Γ_h^3: Re z = $-c_1$,

Γ_h^4: Re z = Im z .

It is well known that for c_1 small enough and C_1 large enough, Γ_h will enclose the (real) spectrum of the negative definite operator D_h^2.

We are going to show that with C a generic constant (independent of h,z),

$$\text{(R)} \qquad \|R_h^z f_h\|_{L_\infty} \leq \frac{C}{|z|} \|f_h\|_{L_\infty}, \quad z \in \Gamma_h, \quad f_h \in S_h.$$

It would then follow that

$$\text{(13)} \qquad \|E_h(t)v_h\|_{L_\infty} \leq C \int_{\Gamma_h^1} h^2 \|v_h\|_{L_\infty} + C \int_{\Gamma_h^2 \cup \Gamma_h^4} \frac{1}{|z|} \|v_h\|_{L_\infty} + C \int_{\Gamma_h^3} \|v_h\|_{L_\infty} \leq C \ln 1/h .$$

<u>Remark 2</u>.

For $t \geq 1$, it is possible to get an estimate Ce^{-kt}.

Similarly, using that $e^{-kt} \leq e^{-1}/(kt)$, for $kt > 0$, we have

$$\text{(14)} \qquad \|\frac{\partial}{\partial t} E_h(t)v_h\|_{L_\infty} \leq C \int_{\Gamma_h^1} e^{-cth^{-2}} \|v_h\|_{L_\infty} + C \int_{\Gamma_h^2 \cup \Gamma_h^4} e^{-t|z|/\sqrt{2}} \|v_h\|_{L_\infty}$$

$$+ C \int_{\Gamma_h^3} e^{-tc_1} \|v_h\|_{L_\infty} \le \frac{C}{t} \|v_h\|_{L_\infty} .$$

The improvement needed for small t is easily obtained. Referring to [3, Lemma 1.3] for the simple proof, we have

$$(15) \qquad \|D_h^2 \chi\|_{L_\infty} \le \frac{C}{h^2} \|\chi\|_{L_\infty} , \text{ for } \chi \in S_h .$$

Thus by (13),

$$(16) \qquad \|\frac{\partial}{\partial t} E_h(t) v_h\|_{L_\infty} = \|D_h^2 E_h(t) v_h\|_{L_\infty} \le \frac{C}{h^2} \|E_h(t) v_h\|_{L_\infty} \le \frac{C}{h^2} \ln 1/h \|v_h\|_{L_\infty} .$$

From (13), (14) and (16) we have the Proposition.

It remains to derive the resolvent estimate (R). Consider first an estimate on Γ_h^1: Letting $g_h = R_h^z f_h$, so that

$$z g_h = f_h + D_h^2 g_h$$

we have by (15),

$$|z| \|g_h\|_{L_\infty} \le \|f_h\|_{L_\infty} + \frac{C}{h^2} \|g_h\|_{L_\infty} .$$

Hence if $|z| = 2Ch^{-2}$ on Γ_h^1, we obtain (R) over this piece.

Consider next Γ_h^3: Here,

$$\|g_h'\|_{L_2}^2 + z\|g_h\|_{L_2}^2 = (f_h, g_h)$$

or,

$$\|g_h'\|_{L_2}^2 \le \|f_h\|_{L_2} \|g_h\|_{L_2} + |z| \|g_h\|_{L_2}^2 .$$

By Poincaré's inequality, $\|g_h\|_{L_2} \le \pi^{-1} \|g_h'\|_{L_2}$ and thus, for $|z|$ small enough, i.e., c_1 small enough,

$$\|g_h'\|_{L_2} \le C\|f_h\|_{L_2} \le C\|f_h\|_{L_\infty} .$$

Since $\|g_h\|_{L_\infty} \le \|g_h'\|_{L_2}$ we have (R) on Γ_h^3.

It remains to show (R) on Γ_h^2 and Γ_h^4. We shall treat Γ_h^4, where $z = y+iy$, $-C_1(\sqrt{2}h^2)^{-1} \le y \le -c_1$. The proof over Γ_h^2 is very analogous.

We need a number of auxiliary results.

Lemma 1.

Let, for $z \in \Gamma_h^4$,

$$(17) \qquad \begin{aligned} -g''+zg &= f, \quad 0 \le x \le 1, \\ g(0) &= g(1) = 0, \end{aligned}$$

so that $g = R^z f$, $R^z = (z - D^2)^{-1}$. Then

$$(18) \qquad \|g''\|_{L_2} + |z|\, \|g\|_{L_2} \le C\|f\|_{L_2}.$$

Proof. We have

$$\|g'\|_{L_2}^2 + z\|g\|_{L_2}^2 = (f,g).$$

Taking imaginary parts,

$$\frac{|z|}{\sqrt{2}} \|g\|_{L_2} \le \|f\|_{L_2}$$

and then the equation (17) itself furnishes the estimate for $\|g''\|_{L_2}$.

Consider next the Green's function $G^{x_0}(x)$ for (17). With $r = \sqrt{z}$ (the branch with $\operatorname{Re} r > 0$, say) we have

$$G^{x_0}(x) = \begin{cases} \dfrac{(e^{rx}-e^{-rx})(e^{rx_0}-e^{r(2-x_0)})}{2r(e^{2r}-1)}, & 0 \le x \le x_0, \\[2ex] \dfrac{(e^{rx_0}-e^{-rx_0})(e^{rx}-e^{r(2-x)})}{2r(e^{2r}-1)}, & x_0 \le x \le 1. \end{cases}$$

One then easily obtains

Lemma 2.

For $z \in \Gamma_h^4$, $x \neq x_0$,

$$\left|\left(\frac{\partial}{\partial x_0}\right)^k G^{x_0}(x)\right| \le C \sqrt{|z|}^{\,k-1} e^{-c\sqrt{|z|}\,|x-x_0|} \qquad \left(c = \cos\frac{3\pi}{8}\right).$$

Since for (17), $g(x_0) = (G^{x_0},f)$ we find without effort

Lemma 3.

For $z \in \Gamma_h^4$,

$$\|R^z f\|_{L_\infty} \leq \frac{C}{|z|} \|f\|_{L_\infty} .$$

We next consider a projection based on (17).

Lemma 4.

Let $z \in \Gamma_h^4$. Let $P_h^z v \in S_h$ be the projection defined by

(19) $$((v-P_h^z v)', \chi') + z(v-P_h^z v, \chi) = 0, \text{ for } \chi \in S_h .$$

Then

(20) $$\|(v-P_h^z v)'\|_{L_2} + \sqrt{|z|}\, \|v-P_h^z v\|_{L_2} \leq C \min_{\chi \in S_h} (\|(v-\chi)'\|_{L_2} + \sqrt{|z|}\, \|v-\chi\|_{L_2})$$

and

(21) $$\|v-P_h^z v\|_{L_2} \leq Ch \min_{\chi \in S_h} (\|(v-\chi)'\|_{L_2} + \sqrt{|z|}\, \|v-\chi\|_{L_2}).$$

Proof. We have for any $\chi \in S_h$,

$$\|(v-P_h^z v)'\|_{L_2}^2 + z\|v-P_h^z v\|_{L_2}^2 = ((v-P_h^z v)', (v-\chi)') + z(v-P_h^z v, v-\chi) .$$

Taking imaginary parts, and using a simple kickback,

(22) $$\frac{|z|}{2\sqrt{2}} \|v-P_h^z v\|_{L_2}^2 \leq \|(v-P_h^z v)'\|_{L_2} \|(v-\chi)'\|_{L_2} + C|z|\, \|v-\chi\|_{L_2}^2 .$$

Therefore,

$$\|(v-P_h^z v)'\|_{L_2}^2 \leq C\|(v-P_h^z v)'\|_{L_2} \|(v-\chi)'\|_{L_2} + C|z|\, \|v-\chi\|_{L_2}^2$$

so that

$$\|(v-P_h^z v)'\|_{L_2}^2 \leq C\|(v-\chi)'\|_{L_2}^2 + C|z|\, \|v-\chi\|_{L_2}^2 .$$

The estimate (20) follows from this and (22).

For (21) we use a duality argument.

$$\|v-P_h^z v\|_{L_2} = \sup_{\|\varphi\|_{L_2}=1} (v-P_h^z v, \varphi) .$$

Let $\varphi = \bar{z}w - w''$, $w(0) = w(1) = 0$. Then for any $\chi \in S_h$,

$$(v-P_h^z v,\varphi) = ((v-P_h^z v)',w') + z(v-P_h^z v,w) = ((v-P_h^z v)',(w-\chi)') + z(v-P_h^z v,w-\chi)$$

$$\leq \|(v-P_h^z v)'\|_{L_2} Ch\|w''\|_{L_2} + |z|\,\|v-P_h^z v\|_{L_2} Ch^2\|w''\|_{L_2} .$$

In the last step a particular suitable χ was chosen and well known approximation properties used.

Using Lemma 1 then (which holds also on Γ_h^2),

$$\|v-P_h^z v\|_{L_2} \leq Ch\|(v-P_h^z v)'\|_{L_2} + |z|h^2\|v-P_h^z v\|_{L_2} .$$

Since $\sqrt{|z|}\,h \leq C$ on Γ_h^4 we obtain (21) by use of (20).

Finally we shall need the following localization of (20). The proof will be given in the Appendix.

Lemma 5.

There exists a positive constant C such that the following holds. Let $\mathcal{J}_1 = [a,b] \cap [0,1]$ with $b-a \equiv d \geq 32h$. Let

$$\mathcal{J}_2 = [0,1] \cap [a-\frac{d}{4}, b+\frac{d}{4}] .$$

Then if $v_h \in S_h$ is such that for $z \in \Gamma_h^4$,

$$((v-v_h)',\chi') + z(v-v_h,\chi) = 0, \text{ for } \chi \in S_h \text{ with support in } \mathcal{J}_2,$$

we have

$$\|(v-v_h)'\|_{L_2(\mathcal{J}_1)} + \sqrt{|z|}\,\|v-v_h\|_{L_2(\mathcal{J}_1)}$$

$$\leq C \min_{\chi \in S_h} \{\|(v-\chi)'\|_{L_2(\mathcal{J}_2)} + \sqrt{|z|}\,\|v-\chi\|_{L_2(\mathcal{J}_2)} + \frac{1}{d}\|v-\chi\|_{L_2(\mathcal{J}_2)}\}$$

$$+ \frac{C}{d}\|v-v_h\|_{L_2(\mathcal{J}_2)} .$$

We are now ready to show that

$$(R)^4 \qquad \|R_h^z f_h\|_{L_\infty} \leq \frac{C}{|z|}\|f_h\|_{L_\infty}, \text{ for } z \in \Gamma_h^4 .$$

Let $g_h = R_h^z f_h$ and $\tilde{g} = R^z f_h$, cf. (17). Let x^0 be an arbitrary point in $[0,1]$, and let I^0 denote a mesh interval I_i containing x^0. Then, by inverse properties,

$$(23) \qquad |g_h(x^0)| \leq Ch^{-1/2}\|g_h\|_{L_2(I^0)} ; \text{ here}$$

(24) $$\|g_h\|_{L_2(I^0)} = \sup_{\substack{\varphi \in \mathcal{C}_0^\infty(I^0) \\ \|\varphi\|_{L_2(I^0)}=1}} (g_h,\varphi) .$$

For each such fixed φ, let $v = R^{\bar{z}}\varphi$. Then, cf. (19) for notation, setting $v_h = P_h^{\bar{z}} v$, since $g_h = P_h^z \tilde{g}$,

$$(g_h,\varphi) = (g_h,-v''+\bar{z}v) = (g_h',v') + z(g_h,v) = (g_h',v_h') + z(g_h,v_h)$$

$$= (\tilde{g}',v_h') + z(\tilde{g},v_h) = (\tilde{g}',(v_h-v)') + z(\tilde{g},v_h-v) + (\tilde{g},\varphi) .$$

Integrating by parts in the first term, and observing that

$$|(\tilde{g},\varphi)| \leq \|\tilde{g}\|_{L_\infty} h^{1/2}\|\varphi\|_{L_2} = \|\tilde{g}\|_{L_\infty} h^{1/2} ,$$

and furthermore that

$$|w(x_i)| \leq C\|\nabla w\|_{L_1(I_i)} + Ch^{-1}\|w\|_{L_1(I_i)} ,$$

we find that

$$|(g_h,\varphi)| \leq C\|\tilde{g}\|_{L_\infty} \{|(v-v_h)'|_{W_1^1,h} + h^{-1}\|(v-v_h)'\|_{L_1} + |z|\,\|v-v_h\|_{L_1} + h^{1/2}\} ,$$

where $$|w'|_{W_1^1,h} = \sum_{i=0}^{N-1} \int_{I_i} |w''|\,dx .$$

By Lemma 3, $\|\tilde{g}\|_{L_\infty} \leq \frac{C}{|z|} \|f_h\|_{L_\infty}$ and so, in view of (23) and (24), $(R)^4$ would follow if we can show that

(25) $$|(v-v_h)'|_{W_1^1,h} + h^{-1}\|(v-v_h)'\|_{L_1} + |z|\,\|v-v_h\|_{L_1} \leq Ch^{1/2} ,$$

for v, v_h (and φ) as above. We shall first obtain

(25)' $$h^{-1}\|(v-v_h)'\|_{L_1} + |z|\,\|v-v_h\|_{L_1} \leq Ch^{1/2} .$$

<u>Remark 3</u>.

A singular perturbation problem $-\varepsilon u''+u = f$, $u(0) = u(1) = 0$, $\varepsilon > 0$, could be treated very analogously. It would follow that the Galerkin solution satisfies

$$\|u-u_h\|_{L_\infty} \leq C \min_{\chi \in S_h} \|u-\chi\|_{L_\infty} .$$

To show (25)', we break $[0,1]$ into a union

$$I^0 \cup I^1 \cup \ldots \cup I^J$$

where

$$I^j = \{x: h2^{j-1} \le |x-x^0| \le h2^j\} \cap [0,1].$$

Set also

$$\tilde{I}^j = I^{j-1} \cup I^j \cup I^{j+1}, \quad \tilde{\tilde{I}}^j = \tilde{I}^{j-1} \cup \tilde{I}^j \cup \tilde{I}^{j+1}.$$

We first consider the L_1 norm in (25)' over the small interval $I^0 \cup \ldots \cup I^5$ around x^0. By Cauchy's inequality, the fact that $h\sqrt{|z|} \le C$, Lemma 4 (which holds also for $P_h^{\bar{z}}$), approximation theory and Lemma 1,

$$\begin{aligned}
&h^{-1}\|(v-v_h)'\|_{L_1(I^0\cup\ldots\cup I^5)} + |z|\,\|v-v_h\|_{L_1(I^0\cup\ldots\cup I^5)} \\
&\le C(h^{-1/2}\|(v-v_h)'\|_{L_2} + |z|h^{1/2}\|v-v_h\|_{L_2}) \\
&\le Ch^{-1/2}(\|(v-v_h)'\|_{L_2} + \sqrt{|z|}\,\|v-v_h\|_{L_2}) \\
&\le Ch^{-1/2}(h\|v''\|_{L_2} + \sqrt{|z|}\,h^2\|v''\|_{L_2}) \le Ch^{1/2}\|\varphi\|_{L_2} = Ch^{1/2}.
\end{aligned} \tag{26}$$

Next, using Lemma 5 (also true for $P_h^{\bar{z}}$),

$$\begin{aligned}
S &\equiv \sum_{j=6}^{J} h^{-1}\|(v-v_h)'\|_{L_1(I^j)} + |z|\,\|v-v_h\|_{L_1(I^j)} \\
&\le \sum_{j=6}^{J} (h2^j)^{1/2}\{h^{-1}\|(v-v_h)'\|_{L_2(I^j)} + |z|\,\|v-v_h\|_{L_2(I^j)}\} \\
&\le \sum_{j=6}^{J} (h2^j)^{1/2}h^{-1}\{\|(v-v_h)'\|_{L_2(I^j)} + \sqrt{|z|}\,\|v-v_h\|_{L_2(I^j)}\} \\
&\le \sum_{j=6}^{J} h^{-1/2}2^{j/2}\{\min_{\chi\in S_h} \|(v-\chi)'\|_{L_2(\tilde{I}^j)} + \sqrt{|z|}\,\|v-\chi\|_{L_2(\tilde{I}^j)} + \frac{1}{h2^j}\|v-\chi\|_{L_2(\tilde{I}^j)}\} \\
&\quad + \sum_{j=6}^{J} h^{-3/2}2^{-j/2}\|v-v_h\|_{L_2} \equiv S_1 + S_2.
\end{aligned}$$

In the first sum S_1, convert to $L_\infty(\tilde{I}^j)$ norms and use approximation theory:

$$S_1 \le C\sum_{j=6}^{J} 2^j\{h + \sqrt{|z|}\,h^2 + h2^{-j}\}\|v''\|_{L_\infty(\tilde{\tilde{I}}^j)} \le C\sum_{j=6}^{J} 2^j h\|v''\|_{L_\infty(\tilde{\tilde{I}}^j)}.$$

By Lemma 2, for $\tilde{\tilde{x}} \in \tilde{\tilde{I}}^j$, since $\|\varphi\|_{L_2} = 1$,

$$v''(\tilde{x}) = \int_{I^0} (G^{\approx x}(x))''\varphi(x)dx \leq C\sqrt{|z|}\, e^{-c\sqrt{|z|}h2^j} h^{1/2} .$$

Hence,

$$(27) \qquad S_1 \leq Ch^{3/2} \sum_j 2^j \sqrt{|z|}\, e^{-c\sqrt{|z|}h2^j} \leq Ch^{1/2} .$$

For S_2, by (21) of Lemma 4, and approximation theory and Lemma 1,

$$(28) \qquad S_2 \leq \sum_j h^{-3/2}\, 2^{-j/2} h(h+\sqrt{|z|}h^2)\|v''\|_{L_2} \leq \Sigma\, h^{1/2}\, 2^{-j/2} \leq Ch^{1/2} .$$

Thus, by (27) and (28), $S \leq Ch^{1/2}$, and with (26), we have proven (25)'.

It remains to show that

$$(29) \qquad |(v-v_h)'|_{W_1^{1,h}} \leq Ch^{1/2} .$$

For any $\chi \in S_h$, we have by inverse properties,

$$|(v-v_h)'|_{W_1^{1,h}} \leq |(v-\chi)'|_{W_1^{1,h}} + |(\chi-v_h)'|_{W_1^{1,h}}$$

$$\leq |(v-\chi)'|_{W_1^{1,h}} + Ch^{-1}\|(\chi-v_h)'\|_{L_1}$$

$$\leq |(v-\chi)'|_{W_1^{1,h}} + Ch^{-1}\|(v-\chi)'\|_{L_1} + Ch^{-1}\|(v-v_h)'\|_{L_1} .$$

Here the last term is already taken care of. The two leading terms reduce to approximation theory. In our case (thanks to our simple setup) we might take χ as a piecewise linear interpolant. Then the first two terms are bounded by $\|v''\|_{L_1}$ which in turn, using Lemma 2, is easily seen to be bounded by $Ch^{1/2}$.

This completes the proof of the resolvent estimate (R), and hence of the Proposition.

APPENDIX.

Proof of Lemma 5.

Assume first that $v \equiv 0$, i.e.,

$$(A.1) \qquad (v_h',\chi') + z(v_h,\chi) = 0, \text{ for } \chi \in S_h \text{ with support in } [a-\tfrac{d}{8}, b+\tfrac{d}{8}] \cap [0,1] .$$

Let ω be infinitely differentiable,

$$\omega \equiv 1 \quad \text{on} \quad \mathcal{J}_1,$$

$$\omega \quad \text{supported in} \quad \tilde{\mathcal{J}}_1 = [a-\tfrac{d}{32},\, b+\tfrac{d}{32}],$$

$$|(\tfrac{\partial}{\partial x})^k \omega| \le C d^{-k}, \quad C \text{ independent of } d .$$

Such a cutoff function ω is easily constructed by scaling in one valid for $d \cong 1$.

Now for any $\chi \in S_h$,

$$\|\omega v_h'\|_{L_2}^2 + z\|\omega v_h\|_{L_2}^2 = (\omega v_h', \omega v_h') + z(\omega^2 v_h, v_h)$$

$$\text{(A.2)} \qquad = ((\omega^2 v_h)', v_h') + z(\omega^2 v_h, v_h) - 2(\omega' v_h, \omega v_h')$$

$$= ((\omega^2 v_h - \chi)', v_h') + z(\omega^2 v_h - \chi, v_h) - 2(\omega' v_h, \omega v_h') .$$

Taking χ to be a suitable local approximant of $\omega^2 v_h$, the right hand side above is bounded by

$$\text{(A.3)} \qquad C\|v_h'\|_{L_2(\tilde{\mathcal{J}}_1)}\, h^{r-1}\|(\tfrac{\partial}{\partial x})^r(\omega^2 v_h)\|_{L_2^h} + C|z|\,\|v_h\|_{L_2(\tilde{\mathcal{J}}_1)}\, h^r\|(\tfrac{\partial}{\partial x})^r(\omega^2 v_h)\|_{L_2^h}$$
$$+ C\|\omega v_h'\|_{L_2} \tfrac{1}{d}\, \|v_h\|_{L_2(\tilde{\mathcal{J}}_1 \cap [0,1])} .$$

Here $\tilde{\tilde{\mathcal{J}}}_1 = [a-\frac{d}{32}-h,\, b+\frac{d}{32}+h] \cap [0,1]$, and L_2^h denotes a piecewise norm. Using Leibnitz rule, that $(\frac{\partial}{\partial x})^r v_h \equiv 0$ on each meshinterval, and $h/d \le \frac{1}{16}$, we find that

$$h^{r-2}\|(\tfrac{\partial}{\partial x})^r(\omega^2 v_h)\|_{L_2^h} \le \tfrac{C}{d^2}\, \|v_h\|_{L_2(\tilde{\tilde{\mathcal{J}}}_1)} + \tfrac{C}{d}\, \|v_h'\|_{L_2(\tilde{\tilde{\mathcal{J}}}_1)} .$$

Inserting this in (A.3) and then in (A.2), we have by use of inverse properties with

$$\tilde{\tilde{\tilde{\mathcal{J}}}}_1 = [a-\tfrac{d}{32}-2h,\, b+\tfrac{d}{32}+2h] \cap [0,1] ,$$

and using also that $h^2|z| \le C$,

$$\big|\, \|\omega v_h'\|_{L_2}^2 + z\|\omega v_h\|_{L_2}^2 \,\big|$$

$$\le C\{h\|v_h'\|_{L_2(\tilde{\tilde{\mathcal{J}}}_1)} + h^2|z|\,\|v_h\|_{L_2(\tilde{\tilde{\mathcal{J}}}_1)}\} \times \{\tfrac{1}{d^2}\, \|v_h\|_{L_2(\tilde{\tilde{\mathcal{J}}}_1)} + \tfrac{1}{d}\, \|v_h'\|_{L_2(\tilde{\tilde{\mathcal{J}}}_1)}\}$$

$$+ \tfrac{C}{d}\, \|\omega v_h'\|_{L_2} \|v_h\|_{L_2(\tilde{\tilde{\tilde{\mathcal{J}}}}_1)} \le \tfrac{C}{d^2}\, \|v_h\|_{L_2(\tilde{\tilde{\tilde{\mathcal{J}}}}_1)}^2 + \tfrac{Ch}{d}\, \|v_h'\|_{L_2(\tilde{\tilde{\tilde{\mathcal{J}}}}_1)}^2$$

$$+ C\frac{h^2|z|}{d}\|v_h\|_{L_2(\tilde{\tilde{\mathcal{J}}}_1)}\|v_h'\|_{L_2(\tilde{\tilde{\mathcal{J}}}_1)} + \frac{C}{d}\|\omega v_h'\|_{L_2}\|v_h\|_{L_2(\tilde{\tilde{\mathcal{J}}}_1)}\,.$$

Thus, considering first imaginary parts and then real parts, and using a kickback argument for the last term,

$$\|\omega v_h'\|_{L_2}^2 + |z|\,\|\omega v_h\|_{L_2}^2 \le \frac{C}{d^2}\|v_h\|_{L_2(\tilde{\tilde{\mathcal{J}}}_1)}^2 + \frac{Ch}{d}\|v_h'\|_{L_2(\tilde{\tilde{\mathcal{J}}}_1)}^2$$

$$+ \frac{Ch^2|z|}{d}\|v_h\|_{L_2(\tilde{\tilde{\mathcal{J}}}_1)}\|v_h'\|_{L_2(\tilde{\tilde{\mathcal{J}}}_1)}\,.$$

Iterating this argument once more, and using inverse properties and that $|z|h^2 \le C$, we find

(A.4) $$\|v_h'\|_{L_2(\mathcal{J}_1)} + \sqrt{|z|}\,\|v_h\|_{L_2(\mathcal{J}_1)} \le \frac{C}{d}\|v_h\|_{L_2(\mathcal{J}_1^*)}$$

where

$$\mathcal{J}_1^* = [a - \tfrac{d}{16}, b + \tfrac{d}{16}] \cap [0,1]\,.$$

Consider now $v - v_h$ as in the statement of the lemma. Let η be a smooth cutoff function with

$$\eta \equiv 1 \quad \text{on} \quad [a - \tfrac{d}{8}, b + \tfrac{d}{8}] \cap [0,1]\,,$$

$$\eta \quad \text{supported in} \quad \mathcal{J}_2\,,$$

$$|(\tfrac{\partial}{\partial x})^k \eta| \le Cd^{-k}\,.$$

Then on $\mathcal{J}_1$,

(A.5) $$v - v_h = (\eta v - P_h^z(\eta v)) + (P_h^z(\eta v) - v_h)\,.$$

Here, by Lemma 4, (taking $\chi = 0$ there for the moment),

(A.6)
$$\|(\eta v - P_h^z(\eta v))'\|_{L_2} + \sqrt{|z|}\,\|\eta v - P_h^z(\eta v)\|_{L_2} \le C\|(\eta v)'\|_{L_2} + C\sqrt{|z|}\,\|\eta v\|_{L_2}$$
$$\le C\|v'\|_{L_2(\mathcal{J}_2)} + C\sqrt{|z|}\,\|v\|_{L_2(\mathcal{J}_2)} + \frac{C}{d}\|v\|_{L_2(\mathcal{J}_2)}\,.$$

The second member of (A.5) is a function in S_h of the type previously considered in (A.1) so that (A.4) can be applied to it. Hence,

$$\|(P_h^z(\eta v) - v_h)'\|_{L_2(\mathcal{J}_1)} + \sqrt{|z|}\,\|P_h^z(\eta v) - v_h\|_{L_2(\mathcal{J}_1)}$$

$$\leq \frac{C}{d}\,\|P_h^z(\eta v) - v_h\|_{L_2(\mathcal{J}_1^*)} \leq \frac{C}{d}\,\|P_h^z(\eta v) - \eta v\|_{L_2(\mathcal{J}_1^*)} + \frac{C}{d}\,\|v-v_h\|_{L_2(\mathcal{J}_2)} \,.$$

Thus, from (A.5) and (A.6),

$$\|(v-v_h)'\|_{L_2(\mathcal{J}_1)} + \sqrt{|z|}\,\|v-v_h\|_{L_2(\mathcal{J}_1)}$$

$$\leq C\{\|v'\|_{L_2(\mathcal{J}_2)} + \sqrt{|z|}\,\|v\|_{L_2(\mathcal{J}_2)} + \frac{C}{d}\,\|v\|_{L_2(\mathcal{J}_1)}\}$$

$$+ \frac{C}{d}\,\|P_h^z(\eta v) - \eta v\|_{L_2} + \frac{C}{d}\,\|v-v_h\|_{L_2(\mathcal{J}_2)} \,.$$

Here by Lemma 4, (21) with $\chi = 0$,

$$\|P_h^z(\eta v) - \eta v\|_{L_2} \leq Ch(\|(\eta v)'\|_{L_2} + \sqrt{|z|}\,\|\eta v\|_{L_2}).$$

We thus obtain Lemma 5 with the minimum on the right replaced with the terms with $\chi = 0$. Writing $v-v_h = (v-\chi) - (v_h-\chi)$ proves the lemma as stated.

REFERENCES

1. J. Descloux, On finite element matrices, SIAM J. Numer. Anal. 9, 1972, 260-265.

2. J. Douglas Jr., T. Dupont and L.B. Wahlbin, The stability in L^q of the L^2-projection into finite element function spaces, Numer. Math. 23, 1975, 193-197.

3. A.H. Schatz, V. Thomée and L.B. Wahlbin, Maximum norm stability and error estimates in parabolic finite element equations, Comm. Pure Appl. Math. 33, 1980, 265-304.

4. A.H. Schatz and L.B. Wahlbin, On the quasi-optimality in L_∞ of the $\mathring{H}'$ projection into finite element spaces, to appear in Math. Comp.

5. V. Thomée and L.B. Wahlbin, to appear.

6. M.F. Wheeler, An optimal L_∞ error estimate for Galerkin approximations to solutions of two-point boundary value problems, SIAM J. Numer. Anal. 10, 1973, 914-917.